© 1999 by Ariel Jerozolimski

PATRICK COCKBURN has been a senior Middle East correspondent for the *Financial Times* and the London *Independent* since 1979. Among the most experienced commentators on Iraq, he was one of the few journalists to remain in Baghdad during the Gulf War. He is currently based in Jerusalem for the *Independent*.

© 1999 by Leslie Cockburn

ANDREW COCKBURN is the author of several books on defense and international affairs. He has also written about the Middle East for *The New Yorker* and coproduced the 1991 PBS documentary on Iraq titled "The War We Left Behind." He lives in Washington, D.C.

ALSO BY ANDREW COCKBURN:

The Threat: Inside the Soviet Military Machine

Dangerous Liaison: The U.S.–Israeli Covert Connection
(with Leslie Cockburn)

One Point Safe: The Leaking Russian Nuclear Arsenal
(with Leslie Cockburn)

ALSO BY PATRICK COCKBURN:

Getting Russia Wrong: The End of Kremlinology

Out of the Ashes

THE RESURRECTION OF SADDAM HUSSEIN

ANDREW COCKBURN AND
PATRICK COCKBURN

HarperPerennial
A Division of HarperCollins*Publishers*

A hardcover edition of this book was published in 1999 by Harper-
Collins Publishers.

HarperCollins books may be purchased for educational, business, or
sales promotional use. For information please write: Special Markets
Department, HarperCollins Publishers Inc., 10 East 53rd Street, New
York, NY 10022.

First HarperPerennial edition published 2000.

Designed by Kris Tobaissen

Map by Paul J. Pugliese

ISBN 0-06-092983-9

00 01 02 03 04 ❖/RRD 10 9 8 7 6 5 4 3 2 1

For Chloe, Henry, Olivia, Alexander, and Charlie

Contents

Acknowledgments

This book has been made possible by the insights, advice, and kindnesses of many people over the years we have covered Iraq. To name them all would be impossible and, in the case of some, unwise. We must however extend special thanks to our editor Terry Karten for her patience, loyalty, and unwavering eye for a redundancy as well as to her indefatigable assistant, Megan Barrett. Our agent, Elizabeth Kaplan, was there when we needed her. Faith Rubenstein performed invaluable service on the research front.

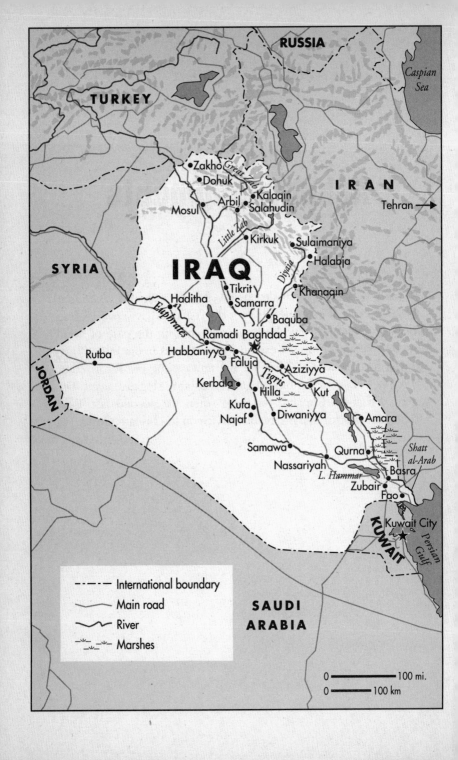

Out of the Ashes

ONE

Saddam at the Abyss

Fifty miles from the capital, returning Iraqi soldiers could already see the black cloud over the blazing al-Dohra oil refinery on the edge of Baghdad. It was early March 1991, and these exhausted men were the remnants of the huge army sent to occupy Kuwait after its conquest by Saddam Hussein the previous year. Now, routed by the United States and its allies, they were in the last stages of a three-hundred-mile flight from the battlefields. They were crowded into taxis, trucks, battered buses—anything on wheels. One group clung desperately to a car transporter.

Soon they were inside the city, only to find it utterly changed. Just six weeks before, the low-lying Iraqi capital on the banks of the Tigris had been a rich modern city, built with the billions of dollars flowing from the third-largest oil reserves in the world. Expressways and overpasses sped traffic past gleaming modern hotels, government buildings, and communications centers. Lavishly equipped hospitals gave the citizens medical care as good as could be found in Europe or the United States. Even the poor were used to eating chicken once a day. Then, beginning at 3:00 A.M. on January 17, precisely targeted bombs and missiles had thrust Baghdad and its 3.5 million inhabitants abruptly back into the third world.

There was no power because all the power stations had been knocked out in the first days of bombing. The people of the city huddled in darkness. The stench of decaying meat hung over the more prosperous districts as steaks in carefully stocked freezers slowly rotted. In the hospitals, doctors trained in the finest medical schools in Europe operated by flashlight.

Like any advanced society, Iraq had been totally dependent on electricity. Water came from the wide Tigris River that flows through the city, pumped and purified by what had been one of the most modern and efficient systems in the world. Now a jury-rigged system brought a muddy brown liquid spluttering out of the taps for just one hour a day. Oil billions had given the city an up-to-date sewage system, but the pumps at the treatment plants had been silent since the power generators had been hit, and every day 15 million gallons of untreated sewage poured into the Tigris.

Few cars moved along the streets and tree-lined avenues because the gas stations had long since exhausted their supplies and al-Dohra, along with all other Iraqi refineries, had been smashed in the bombing. In the sparse traffic, black smoke poured from the exhausts of some vehicles, a symptom of watered-down gas available on the black market at a hundred times the prewar price.

Familiar landmarks lay in ruins, like the handsome Jumhuriya Bridge across the Tigris in the city center, now trisected by allied bombs. Surviving bridges had old sacking draped over the sides and little saplings tied to the railings, a vain effort to deceive the computers and laser-targeting systems of the enemy weapons. Symbols of authority, like the Ministry of Justice, at first glance seemingly untouched, were empty shells, their insides gutted by high explosives. The phones had stopped working when two laser-guided bombs had hit the communications center across from the Mansour Melia Hotel and melted the satellite dishes on the roof, isolating Iraqis from the outside world and each other.

The air was full of smoke from the burning refinery and from piles of tires set alight during the war to confuse allied warplanes. The restaurants on Sadoun Street were shuttered and empty, replaced by curbside cooking fires fueled by branches torn from trees by the bombs. Over everything there hung the yellow haze of a winter fog.

Somewhere beneath the gloom was the man who had caused the disaster, President Saddam Hussein, his thoughts and actions, even his whereabouts in those dramatic days, a mystery to his people and to the outside world.

Physically, he had changed since the war had begun. In the months of crisis between his invasion of Kuwait on August 2, 1990, and the start of the United States–led counterattack in January 1991, the Iraqi leader had played to a global audience. Sleek in the beautiful silk suits created by his Armenian tailor, Saddam had sat in his palaces declaiming to visiting statesmen and journalists on the justice of that invasion, defying the international coalition that was building up its forces to oust him.

Now the president of Iraq moved about his capital like a man on the run. Like the rest of the high command, he had been careful to stay out of the underground command bunkers built for the war against the Iranians in the 1980s. He had known that the Americans would carefully target these places and that their bombs could—and did—penetrate the thickest concrete. The bombing had stopped, but still he was sleeping in a different house every few nights, staying mainly in the middle-class al-Tafiya district of the city, quiet because many of its inhabitants had fled Baghdad.

Once upon a time, Saddam had sought to confuse potential assassins about his movements by deploying whole fleets of identical Mercedeses, choosing the convoy he would use only at the last minute and dispatching the others in different directions as a distraction. These days Saddam drove only in cheap, inconspicuous cars, accompanied by a single bodyguard—a colonel who himself wore no insignia of rank. The few trusted aides and intimates he visited saw a shrunken figure. He had lost as much as forty pounds in the first month of the war. Now the olive-green uniform of his ruling Baath Party hung ever more loosely on him. "I don't know what God will bring tomorrow," he remarked despairingly to one of his intelligence chiefs.

Officially, his government was in denial, issuing statements that the defeat of his army in Kuwait had been a historic victory, that the occupation of that little oil-rich kingdom had been justified, even hinting that Iraq would try again. The few remaining foreign journalists camped on the lower floors of the al-Rashid Hotel (the elevators had long since stopped running) found Ministry of Information censors still routinely changing the phrase "defeat of the Iraqi army in the south" to read "the fate of the Iraqi army in the south" even as Iraqi generals were meekly accepting conditions laid down by the victorious allies.

To the few trusted aides permitted in his presence, the dictator exhibited a greater sense of reality. One of these was a stocky forty-four-year-old general, the chief of military intelligence, Wafiq al-Samarrai, who, like many other ranking servants of the regime, sported a mustache trimmed in the style of his leader. He had made his reputation during the bitter eight-year war with Iran. Saddam valued his professional judgment and had been visiting his emergency headquarters almost every day since the Americans had started bombing Baghdad. (Anticipating that it would be a target, al-Samarrai evacuated his prewar command post days before it was duly crushed by bombs.) On the day after the allied armies began to sweep, almost unopposed, through Kuwait, Saddam made a rare though roundabout confession of error. "In two hundred years," he remarked to al-Samarrai, "nobody will realize that this was a wrong estimate about what would happen."

"This" had been Saddam Hussein's great gamble in August 1990, that he could surprise the world by seizing the little oil emirate of Kuwait on his southern border and get away with it. The gamble had failed, just as his bet a decade earlier that he could invade his neighbor Iran, then in postrevolutionary chaos, had landed him in a bloody eight-year stalemate. The war against Ayatollah Khomeini had at least ultimately garnered him a partial victory, a de facto alliance with the United States, and the strongest military forces in the Persian Gulf. But the Iran-Iraq war had also cost the lives of hundreds of thousands of Iraqis and, more important for Saddam, had saddled him with $80 billion in debts. Kuwait had been a wager that he could refill his coffers and secure a whip hand over the world's most important oil-producing region, but he had not expected the consequences of losing to be so terrible.

The invasion of Kuwait had been his idea alone. At first it seemed a brilliant success. Saddam's elite divisions had overrun the country in hours, sending the Kuwaiti royal family fleeing over their southern border into Saudi Arabia. The United States and the rest of the world had been caught entirely off guard. As his Republican Guards had massed on the Kuwaiti border at the end of July 1990, the consensus of opinion among those watching his moves had been that he would at worst merely seize part of the northern Kuwaiti oil field and possibly two disputed offshore islands. Later, Deputy Prime Minister Tariq Aziz told an interviewer that this limited invasion had indeed been the original plan. At the last minute, Saddam eschewed this cautious approach and went all the way.

Saddam has often been prone to sudden, unpredictable gambits. At a high-level meeting in September 1979, soon after he seized total power in Iraq, he even delivered a brief homily on the utility of such tactics as a political principle. "What is politics?" the recently installed president asked rhetorically in his slightly shrill voice. "Politics is when you say you are going to do one thing while intending to do another. Then you do neither what you said or what you intended." That way, he suggested, no one could predict what you were going to do.

Along with this taste for sudden rolls of the dice, there was a strong element of fatalism in the Iraqi leader. He once told King Hussein of Jordan that ever since his narrow escape after trying and failing to assassinate Iraqi president Abd al-Karim Qassim in 1959, he had felt that every extra day of life was a gift from God. "I consider myself to have died then," he declared. He acknowledged only one greater power. On a visit to Kuwait after his conquest, he talked to thirty of his senior commanders. A tape of the meeting, later smuggled out of Iraq by a dissident, records him describing the invasion as part of his messianic mission. "This decision to invade Kuwait we received almost ready-made from God," he says. "Our role is simply to carry it out." The audience response was limited to shouts of "God is great."

If Deputy Prime Minister Tariq Aziz, Saddam's perennial voice to the outside world, is to be believed, he did at least try to point out to the leader what the consequences of the invasion might be. In late March 1991, Aziz met with an old friend, the Jordanian politician Zeid Rifai, for the first time since the invasion of Kuwait. "What did you people think you were doing?" asked Rifai. "Didn't you realize what would happen if you seized Kuwait?"

"The leadership made some mistakes," mumbled a slightly crest-fallen Aziz, a dangerous enough admission to anyone but an old friend. They both knew who "the leadership" was.

"Well, why didn't you try and talk him out of it?"

"I did," Aziz explained. Just before the Iraqi army crossed the border, Saddam had finally revealed the full dimensions of the plan to members of his cabinet, who were unaware that the limited incursion originally planned had been drastically enlarged. Aziz chose an indirect way to point out to the boss that this could be a perilous undertaking. "I said, 'The Americans may come to Saudi Arabia and counterattack. Why don't we go all the way and take Saudi Arabia too?'" In suggesting an even bigger gamble, he hoped that his master might reflect on the hazards of the invasion plan. But Saddam took it straight, gently chiding Aziz for his impetuosity.

"In that circle, the safest course is always to be ten percent more hawkish than the chief," says one veteran Russian diplomat long stationed in Baghdad. "You stay out of trouble that way."

There was no one left to stand up to Saddam. In 1986, when Iraq was on the verge of defeat in its war with Iran, the professional army generals had secured some leeway in directing military operations. As soon as they had finally won a narrow victory with the active help of the U.S. Navy in the Persian Gulf, Saddam got rid of them. Some were executed, others retired. Defense Minister Adnan Khairallah Tulfah, Saddam's first cousin but widely liked and respected in the army, died in 1989 in a helicopter crash during a sandstorm. It is a measure of the violence of Iraqi politics that everybody in Baghdad assumed that Saddam had arranged for the helicopter to be sabotaged, though the storm was violent enough to blow the roof off the headquarters of military intelligence. Queried by a foreign interviewer about his purges of the military during the Iran-Iraq war, Saddam was less than reassuring: "Only two divisional commanders and the head of a mechanized unit have been executed. That's quite normal in war."

Once installed in Kuwait, Saddam utterly failed to appreciate the game he had started, and continued to overplay his hand. At the end of August, he met Yasser Arafat, the Palestinian leader, and Abu Iyad, Arafat's chief lieutenant, who were in Baghdad in a vain attempt to mediate. "If I make a peace proposal," Saddam told the

Palestinians, "then I'm the one who will have to make concessions. If the others propose one, then I can obtain concessions."

But President George Bush, steadily building up military strength in Saudi Arabia, had less and less reason to compromise. Saddam wholly underestimated the strength of the coalition that was about to attack him. Just before the war, he appealed to Arab and Muslim solidarity by, among other measures, redesigning the Iraqi flag to include the Islamic rallying cry "Allah Akbar"—"God is great." Iraq did enjoy popular sympathy in the Arab world, but no powerful friends. Saddam had invaded Kuwait at the very moment that the Soviet Union, Iraq's old ally, had gone into terminal decline. He had failed to understand the military superiority of the American-led alliance, entertaining the fantasy that if there was fighting, his troops could withstand bombing from the air and could inflict heavy casualties on any allied ground assault. In the secret meeting with his commanders in Kuwait before the war, he told them that during allied air raids they should "stay motionless underground just a little time. If you do this, their [bombing] will be in vain. . . . On the ground the battle will be another story."

The truth seems to have dawned on Saddam that war was inevitable only after a fruitless meeting between Tariq Aziz and Secretary of State James Baker five days before the war. Even then, little was done to prepare ordinary Iraqis for war. When allied planes approached Baghdad at 2:58 on the morning of January 17, their pilots were astonished to discover that there was no blackout and that the Iraqi capital was "lit up like Las Vegas." Government ministries were floodlit.

Some of the population still trusted their leader to avoid war. Trainers at the racetrack in Mansour, a fashionable district full of foreign embassies in the center of Baghdad, were still walking racehorses on the afternoon before the first bomb attacks. No one had any delusions as to what war would mean if it did come. Despite Saddam's bombast about "the mother of all battles," the feeling in the streets was resigned, with few expectations other than the inevitability of defeat. Pro-government rallies in Baghdad just before the war started consisted entirely of schoolchildren assembled by officials of the ruling Baath Party. The largest public meeting in the city in the days before the bombing turned out to be a gathering of pigeon-racing enthusiasts. Nor were the Iraqis ill-informed about the approaching war. There was little on Iraqi radio or television, but people spent hours listening to foreign radio stations in Arabic, switching from the BBC to Monte Carlo to Voice of America. "Our main hobby is listening to the radio," one Iraqi told us at the time. In the days before the bombing, as many as 1 million out of 3.5 million people in Baghdad left the city. They feared that if Iraq fired a Scud with a chemical or biological warhead at Tel Aviv, Israel would respond with a nuclear strike.

At the start of the bombing, an old man drinking tea in a dilapidated café near Nasr Square explained what he thought, using a double-edged story. He repeated the old Koranic tale of how once "the Abyssinians brought elephants to conquer Mecca. At first the bedouin warriors were dismayed by the strange beast, but God sent birds to Mecca who dropped stones on the elephants and killed them." Saddam himself had recently told the same story, adding that he had only just learned the significant fact that the elephant was the symbol of President Bush's Republican Party. But unlike the Iraqi leader, the old man told the story with exaggerated gestures, to the sound of giggles from the others in the cafe. Not a dissident word was expressed, but the message seemed clear: Unless God could come up with magical birds, Iraq had no hope against the allied elephants.

The mood among the soldiers was scarcely more optimistic. In the last days of peace, Saddam visited the trenches in Kuwait and talked to soldiers. They were, plainly, terrified by his presence. The conversations were full of agonizing pauses.

"Where are you from?" he asked one.

"Sulaimaniya, in Kurdistan."

"How are the people in Sulaimaniya?"

"They support you."

A general who later fled to exile in England explained to us that the low morale in the army in Kuwait at the start of the fighting was not because of superior allied weapons. "We knew all about these weapons. We were all circulated with a newsletter about such developments." They simply thought they had been led into an insane enterprise. "We didn't expect a war. We thought it was all a political maneuver."

If Saddam was aware of his subjects' views, he paid little attention. He was under no illusion that they actually liked him. Long before, soon after the 1968 coup that had put his Baath Party in power, Saddam had spoken with a family who had come to complain that one of them had been unjustly executed. "Do not think you will get revenge," he had said then. "If you ever have the chance, by the time you get to us there will not be a sliver of flesh left on our bodies." He meant that there would be too many others waiting in line to tear him and his associates apart.

Since that time, Saddam had eliminated all potential rivals while his host of secret police and intelligence agencies visited immediate and terrible punishment on anyone manifesting, the moment they were detected, the slightest signs of political discontent. He came from Ouija ("the crooked one"), a typical Iraqi village of flat-roofed brick houses, just outside the decayed textile town of Tikrit, perched on the bank of the Tigris a hundred miles north of Baghdad. Even before Saddam, the Tikritis were known for their violence. A British official writing soon after the First World War spoke of "their ancient reputation for savagery and brutality." He favored razing the town to the ground. Saddam's family belonged to the Bejat clan, who were in turn linked to the tribes in and around Tikrit. Their members formed the core of Saddam's regime and consequently expected little mercy if he fell. Tikritis like the Saddam family belonged to the Sunni branch of Islam. Sunnis, who live mostly in the center and north of the country, make up only 20 percent of the total Iraqi population, but they dominated the upper ranks of the army and the administration, as they had since the days when Iraq was part of the Ottoman Empire.

The majority of Iraqis were Shia Muslims, like the Iranians across the eastern border. Concentrated in Baghdad and on the great flat plain of southern Iraq that stretches all the way to Kuwait and Saudi Arabia, they provided much of the rank and file for the army but were seldom allowed to rise to positions of influence in any Iraqi regime. Since the Baathists had seized control of Iraq, the power of both the political parties they supported and the traditional Shia tribal sheikhs had been whittled away. If the Shia showed loyalty to any figures outside the government, it was to their religious leaders. Saddam had instituted a thorough purge of such figures in the early stages of his confrontation with Iran. The survivors had remained quiet.

The Kurds in the mountainous north had always been more of a problem than the Shia. Non-Arab Sunni Muslims, the Kurds of Iraq saw themselves as a separate community and had resented rule from Baghdad even in the days when the British held sway there. In the early 1970s, backed by the United States and the shah of Iran, they had launched a fierce insurgency that was defeated only when they were betrayed by their foreign friends. During the Iranian war of the 1980s, some of their leaders had again risen in rebellion and Saddam had retaliated by showering poison gas on Kurdish civilians and by ordering a program of mass executions that killed as many as two hundred thousand Kurds. In addition to this holocaust, the Iraqi leader had wiped four thousand Kurdish villages off the map, herding their inhabitants into cities and refugee camps under the ever-suspicious eyes of his secret police. In the months of crisis that followed Saddam's invasion of Kuwait, the principal Kurdish leaders, Massoud Barzani and Jalal Talabani, seemed to have learned their lesson, pledging neutrality to the Iraqi leader in his confrontation with the allies.

Prior to the invasion of Kuwait and the threat to world oil supplies, Saddam's murderous regime evoked few complaints in the outside world. Even when he took to gassing his Kurdish subjects, governments in Washington, London, and other Western capitals stayed mute, grateful that he was fighting the Islamic Republic of Iran. A strictly enforced rule, laid down after a meeting between Jalal Talabani and a mid-level State Department official in 1988 had drawn an angry protest from Baghdad, forbade any U.S. government official from meeting with any of the exiled Iraqi opposition groups. In 1991, as the United States and other members of the coalition began bombing Iraqi cities, there was no move to rouse the people of Iraq against their dictator. The universal assumption abroad was that in such a viciously efficient police state, where even spilling coffee on the leader's picture in a newspaper could bring swift punishment, there was no prospect of any challenge to the regime from below.

Then, on February 15, a full month into the war, President George Bush suddenly spoke directly to ordinary Iraqis. Twice that day, at the

White House and at a missile plant in Massachusetts, he repeated a carefully phrased call for revolt, calling on "the Iraqi military and the Iraqi people to take matters into their own hands and force Saddam Hussein, the dictator, to step aside." The appeal had been conceived of as an incitement to the Iraqi military to stage a coup, and the "Iraqi people" had been included only as an afterthought, but the effects were far-reaching. The president's unequivocal words were broadcast on every international channel that reached Iraq, and millions of Iraqis heard the call. It seemed to them that Bush, Saddam's enemy, whose planes were bombing the country at will, had asked them to join his invincible coalition.

The army in Kuwait, manned largely by Shia and Kurdish conscripts, was already unwilling to die for Saddam. Once they had realized that the "political maneuver" had failed, they had begun to vote with their feet. Captain Azad Shirwan, an intelligence officer with a tank brigade stationed on the front lines in Kuwait, remembers that by the time the allied ground offensive started on February 24, most of the men in his unit had disappeared. "In our brigade, positions were mostly defended by officers, because the private soldiers had deserted." When Saddam suddenly ordered a general withdrawal from Kuwait the day after the allied ground offensive began, the disintegration became total.

The disappearance of the Iraqi troops bemused the allied generals, who had, in any case, vastly exaggerated the strength of their enemy. "What really amazes me is the lack of bodies," exclaimed General Charles Horner, the U.S. Air Force commander. "There weren't a lot of dead people around. I think a lot of Iraqis just left." Later, the U.S. government deliberately avoided quantifying the enemy dead for fear that a huge number would serve as useful propaganda for Saddam. In fact, the available evidence suggests that the number of Iraqi casualties was extraordinarily low. "We didn't lose a single officer over the rank of brigadier," says General al-Samarrai, who, as head of military intelligence, was in a position to know.

Casualties among the lower ranks were also light. In one small village, Tulaiha, just off the main road between Baghdad and Kut, 150 men were called up to the army during the Gulf War. Hassan Hamzi, the *mukhtar,* or village leader, insisted that none of them was killed or wounded. The only casualties were two men captured. This compared with thirty dead and eleven prisoners from Tulaiha during the Iran-Iraq war. While Iraq lost 2,100 tanks in Kuwait, U.S. damage-assessment teams found that only 10 percent had been destroyed in battle. The rest had been abandoned.

In the last few days of February, hundreds of thousands of angry soldiers were streaming out of Kuwait, bitter at Saddam Hussein for starting a war they could not win. Hard on the heels of the disappearing enemy, the allied armies swept through Kuwait and across the border into Iraq itself. Saddam thought they might be coming for him. In the final days of the war, he turned up at military intelligence headquarters with his powerful and sinister private secretary, Abed Hamid Mahmoud. "Abed Hamid thinks the allies are coming to Baghdad," he said to General al-Samarrai. "What do you think?" The general disagreed. On February 28, George Bush proved him right by calling a cease-fire; the allied onslaught halted in its tracks. Though his Kuwaiti adventure had turned into a colossal disaster, Saddam now thought the crisis had ended. "After the cease-fire, he thought everything was finished," explains al-Samarrai. It was, in fact, just beginning.

When he first heard the news that Iraqis themselves had risen in revolt, General al-Samarrai was at the emergency headquarters in which he had spent the war, unmolested by the American bombers. The tidings came in a phone call from Basra, far to the south and near the Kuwaiti border. An army general, Hamid Shakar, had been driving to Baghdad with one bodyguard when unknown rebels had attacked and killed him near a paper mill thirty miles north of Basra. al-Samarrai contacted Saddam, who rushed to the headquarters. He had just arrived, visibly worried, when the phone rang again. al-Samarrai picked it up and recognized the voice of General Nizar Khazraji, the commander of the entire southwest of the country, with his headquarters in Nassariyah, two hundred miles from Baghdad.

"The rebels are trying to attack us," Khazraji shouted. To convince Baghdad of the seriousness of his situation, he held up the phone, saying, "Don't you hear the sound of the bullets?" The connection was poor and al-Samarrai could hear nothing over the crackling. The besieged commander pleaded for a helicopter to rescue him.

"I told Saddam, who was still sitting in my headquarters, what was happening in Nassariya and he ordered a helicopter to rescue Khazraji," says al-Samarrai. But the army in the south was disintegrating fast. Shia conscripts were turning on any representative of Saddam's government, including senior officers. The commander of the Iraqi helicopter force said that nothing could be done: "We don't have any helicopters in the area." Soon afterward all contact with the besieged headquarters was lost. Later Saddam and al-Samarrai heard it had been stormed by the rebels and Khazraji severely wounded.

Fanned by the rage of the soldiers streaming out of Kuwait, the revolt spread with the speed of a whirlwind through the cities and towns of the south. Saddam was now staring into the face of disaster.

"We were anxious to withdraw, to end the mad adventure, when Saddam announced withdrawal within twenty-four hours—though without any formal agreement to ensure the safety of the retreating forces," one officer recounted later. "We understood that he wanted the allies to wipe us out: He had already withdrawn the Republican Guard to safety. We had to desert our tanks and vehicles to avoid aerial attacks. We walked a hundred kilometers toward the Iraqi territories, hungry, thirsty, and exhausted." Finally they arrived at the first little town inside their own border. "In Zubair we decided to put an end to Saddam and his regime. We shot at his posters. Hundreds of retreating soldiers came to the city and joined the revolt; by the afternoon, there were thousands of us. Civilians supported us and demonstrations started. We attacked the party building and the security services headquarters."

At 3:00 A.M. on the first of March, the storm reached Basra, the ancient, sprawling city at the junction of the Tigris and Euphrates rivers where in happier days vacationers from teetotal Kuwait had thronged the hotels and nightclubs in search of a bottle of Johnnie Walker Black Label. A single tank gunner expressed his anger at the debacle by firing a round through a portrait of Saddam Hussein, one of the tens of thousands of such pictures that gazed out on every street throughout the country. The soldiers around him applauded his spontaneous act. Within hours, the iron control of Saddam and the Baath Party had been violently cast aside. For the millions of Iraqis who had suddenly found their voices after years of terrified silence, it was the "intifada"—the uprising.

The first that Dr. Walid al-Rawi, the administrator of Basra Teaching Hospital, knew about the uprising was when a policeman visited him to say that incidents were starting in small towns and villages around Basra. "Later that day, a band of fifty rebels came to the hospital and took away three patients who were security men, one of whom they shot on the hospital grounds." As in cities elsewhere in the south, the Baath Party offices were the first to come under attack. Mohammed Kassim, the manager of the Basra Tower Hotel, later told us that on the first day of the uprising, armed men came to his hotel. "They asked if there were any Baathists staying, or any alcohol," he recalled. "I told them no and they went away." The manager of the nearby Sheraton was less persuasive, or perhaps the rebels were less compliant. They set fire to the top stories of the hotel, burning nineteen rooms.

Rampaging through the city, the rebels made a chilling discovery. Beneath the BATA shoe company premises opposite the mayor's office, they found a secret underground prison. Some of the hundreds of prisoners had been shut off from the world so long that they shouted "Down with al-Bakr" as they were released and led into the open air. They believed that the president of Iraq was still Ahmed Hassan al-Bakr, who had been replaced by Saddam Hussein in 1979.

Within days, the intifada had spread to the holy cities of Kerbala, Najaf, and Kufa, the heartland of the Shia religious tradition to which 55 percent of Iraqis belong. Thirteen hundred years before, the men the Shia regarded as the Prophet's true heirs, Imam Ali and his sons, Hussein and Abbas, had been martyred here, and their shrines are the focus of adulation from the 130 million devotees of the Shia faith around the world.

In Najaf, where for a thousand years pilgrims had flocked to the great shrine, its golden dome rising above the low brick houses of the city, the allied bombing had killed thirty-five people. Thirteen members of the al-Habubi family had been crushed by a stick of bombs that had missed a nearby electricity substation and had turned their house into a gray concrete sandwich, the floors collapsing on each other. The rebels said such horrors only underlined the government's inability to protect its people from air attack. At the funeral of a religious notable, Yusuf al-Hakim, on February 14, the crowd chanted against Saddam. By the time the angry rabble of military deserters started straggling into the city in the first two days of March, the government's authority was already fragile.

Brigadier Ali, a professional officer, was among the returning crowd. Born in Najaf, he and many other deserters from the city arrived home on March 2 after being "chased like rats" out of Kuwait. "The streets were full of deserters. All structure in the army was lost. Everybody was their own boss. News was spreading that someone had shot at Saddam's portrait in Basra."

The next day, Ali heard there was to be a demonstration in Imam Ali Square, four hundred yards from the great shrine at the center of the city. "At first there were about a hundred people, many of them army officers from Najaf who had deserted. The security forces were well informed and were there as well. The demonstrators started shouting: 'Saddam, keep your hands off. The people of Najaf don't want you.'"

The security men opened fire, further infuriating the demonstrators. Only a few of them were armed, but they threw themselves on the detested but hitherto invulnerable officials. Catching one important local Baath Party functionary, they hacked him to death with knives. Now more people had flooded into the area, drawn by the sound of the shooting. As the security men continued to fire, the demonstrators ran into the warren of alleys and small shops between the square and the shrine. The security forces dashed after them, but the gunfire echoed and reverberated off the walls of the ancient market and they became confused, lost heart, and retreated to their headquarters. It had been no more than twenty minutes or half an hour since the first shouts denouncing Saddam, and the crowd of teenagers and young men in their twenties now controlled the center of the city. Their morale soared.

In a few hours, the newly confident crowd took over the shrine of Imam Ali itself, a golden mosque at the center of a courtyard surrounded by rooms for pilgrims. Unlike the rest of Najaf, the shrine had power from a generator and the demonstrators commandeered its loudspeakers, normally used to call people to prayer, to broadcast simple slogans—"Seek out the criminals"—and urge a final attack on the security forces.

In the evening, the insurgents fought their way into the girls' school used as a local headquarters by the Amn al-Khass, one of the many Iraqi secret police services, and killed eight or nine people there. They were increasingly well armed, having seized submachine guns stockpiled by the government in schools to arm people in case the allies landed from the air. The headquarters of the Quds division of the Republican Guard was just outside the city, but all its combat brigades had been sent to the front and the only garrison consisted of some administrative personnel. These did not resist when officers among the rebels commandeered 82-millimeter mortars and used them to bombard the Baath Party headquarters. "Abdel Amir Jaithoum, my old headmaster, was killed there," recalls Brigadier Ali without regret. "So too was Najim Mizhir, who was the only Baath leader in the city who actually came from Najaf and was quite liked, though he shot a demonstrator." Other Baath Party members fled for their lives through the city's immense cemeteries, filled with the graves of devout Shia from around the world. By early morning on March 4, the rebels ruled Najaf; within a day, they also held Kerbala, Kufa, and the entire middle Euphrates area.

As Saddam's rule collapsed across southern Iraq, he was assailed by a fresh crisis at his rear. News arrived from the north that the Kurds had also risen.

Unlike the spontaneous and leaderless fury of the southern intifada, the Kurdish revolt was planned. While publicly refusing to take advantage of Saddam's confrontation with the international coalition, the Kurdish leaders had begun planting the seeds of an insurrection well before the end of the war. Massoud Barzani, the small and boyish-faced tribal chief who headed the Kurdistan Democratic Party (the KDP) led by his father years before, had forged an alliance with the other principal commander, Jalal Talabani, the barrel-chested and garrulous leader of the Patriotic Union of Kurdistan (the PUK). They controlled guerrilla forces, turbanned Peshmerga whose fathers and grandfathers had fought mountain campaigns for half a century against regimes in Baghdad. Before and during the war, agents dispatched by Barzani and Talabani had secretly infiltrated the Jash, a Kurdish militia force recruited by Saddam, in preparation for the moment when their enemy might be weakened enough by the allies for them to strike. As in the country of the Shia, George Bush's call to the Iraqi people had resonated with the Kurds, and they had tentatively scheduled the start of their revolt for the middle of March.

The explosion came sooner than that, catching the leadership by surprise. On March 5, in the small mountain town of Rania, police tried to round up some of the army deserters who had arrived home from the debacle in Kuwait. The local Jash, already suborned by agents of the underground resistance, reacted by seizing control of the town. Within hours the revolt had spread across the sharp crags and winding, narrow canyons of the Kurdish mountains to Sulaimaniya, the provincial capital close to the Iranian border. Here, after two days of hard fighting, the rebels captured the stone fortress that served as the long-dreaded Central Security Headquarters, potent symbol and instrument of the regime. Behind the imposing front entrance, decorated with a giant all-seeing metal eye, they found a medieval warren of torture chambers, equipped with metal hooks, piano wire, and other devices, and smeared with blood. In some rooms, the insurgents discovered freshly strangled women and children. In one, a human ear was nailed to the wall. As in Basra, some of the prisoners had been sealed in underground cells for more than a decade. The outraged crowd fell on the four hundred Baath Party members, intelligence officers, and secret police agents who had holed up in the security headquarters when the revolt began, and massacred them all.

The careful plans of the leadership were swept away as the northern intifada swept across the cities of the mountains and down onto the plains. Two weeks after the first outbreak of rebellion in Rania, the Kurdish Peshmerga guerrillas captured the vital oil center of Kirkuk, only a few hours' drive from Baghdad. "One second of this day is worth all the wealth in the world," cried an exultant Massoud Barzani. Everywhere people celebrated the man they regarded as their ultimate inspiration with the honorific title "Haji." "Haji Bush," they cried to the few Western correspondents who made their way into liberated Kurdistan at the end of March.

Saddam had now lost control of fourteen of Iraq's eighteen provinces. Baghdad itself remained quiet, but government officials were already showing a readiness to desert the sinking ship. Rumors spread that Saddam had fled the country. In Washington and London, allied officials relaxed in the comforting assumption that no leader could survive such disasters. They were wrong.

The uprisings had taken the rest of the world, as well as Saddam Hussein, completely by surprise. Years before, during the Iran-Iraq war, his exiled opponents had miscalculated the strength of the Iraqi patriotism that he was able to enlist on his side after Iranian forces entered Iraq in 1982. In the crisis after the invasion of Kuwait, exiles made the opposite mistake, underestimating popular anger against Saddam Hussein. When rebellion swept through southern Iraq, the opposition had no organization in the cities capable of directing events. In the town of Hillah, for example, only sixty-six miles from Baghdad, a rebel officer proposed taking the six tanks under his command and leading them to the capital. "The way to Baghdad is open," he cried, but his fellow deserters preferred to concentrate on lynching local Baath officials. In Najaf and elsewhere, euphoria at the overthrow of the regime was followed by anarchy. "At first we were a little crazy," recalls Hameed, a schoolteacher, about these first days in Najaf. "We believed even the traffic lights represented Saddam Hussein, so we wrecked them." Three days after the mob drove out Saddam's forces, there were still dead bodies lying in the streets.

There was only one man in the city with authority, albeit of a spiritual nature. The ninety-one-year-old Grand Ayatollah Abu al-Qassim al-Khoie was the most universally respected cleric in all of Shia Iraq. He was the grand *Marja,* the Shia equivalent of pope. Born in northwestern Iran, like many other clerics from outside Iraq (Ayatollah Khomeini himself had lived there for sixteen years) he had long lived and taught in Najaf. Unlike Khomeini, al-Khoie was opposed to the Shia clergy taking power themselves. To the students and disciples gathered around him at the Green Mosque in the great shrine he had always preached that involvement in politics corrupted religion.

On March 6, the frail but venerated grand ayatollah issued a *fatwa,* a religious decree, telling the people: "You are obligated to protect people's property, and honor, likewise, all public institutions, for they are the property of all." He urged the burial of bodies, though without success.

The mood in Najaf was euphoric but confused. "Nobody knew what was going on, but they knew that the city was in our hands," says Sayid Majid al-Khoie, the second son of the grand ayatollah. The night after the rebels took the city he visited the shrine of Imam Hussein and wrote in his diary what the people were saying. "Iraq is finished," said one man, "the Western armies are in Basra and Samawa" (on the Euphrates). Others were saying, "Kerbala and Najaf are in our hands. Let us go on to Baghdad." People eagerly repeated the rumor that Saddam had left Iraq.

That same day, army officers, encouraged by the ayatollah, formed a committee, but they could not impose discipline on the young men who had led the first demonstrations and now ruled the streets. They could not even take advantage of events that seemed to play into their hands. The commander of the battalion spearheading the government counterattack on Kerbala shot his chief security officer in the head and changed sides. "But the committee could not keep his unit together," laments one of its members. "We had to tell the men to change into their dishdashes [civilian robes] and go home."

In towns along the Euphrates close to the Iranian border, someone did lay claim to leadership. Mohammed Baqir al-Hakim, the scion of a revered Shia religious family, had rallied to the Iranian side in the war of the 1980s. Now, from a town just across the border from Basra, he commanded the Supreme Council for the Islamic Revolution in Iraq (SCIRI), and soon, on walls in Basra and in towns, like Amara, close to the border, posters of al-Hakim and the late Iranian leader Ayatollah Khomeini himself began to appear. Announcements in al-Hakim's name claimed full authority over the rebellion: "No action outside this context is allowed; all parties working from Iranian territories should also obey al-Hakim's orders; no party is allowed to recruit volunteers; no ideas except the rightful Islamic ones should be disseminated."

Nothing was more likely to isolate the rebels. The prospect of an

Islamic revolution frightened large numbers of Iraqis, such as Sunni Muslims, Kurds, Christians, secular Iraqis, and anyone associated with the Baath Party. Nor were the United States and its allies likely to be reassured by such slogans. So convenient was it for Saddam Hussein for the uprising to be identified with Iran and militant Islam that some Iraqi opposition leaders were quick to believe that he had planted evidence of Iranian involvement. "He sent his own Mukhabarat [secret police] to the south with pictures of Khomeinei," insists Saad Jabr, a veteran of the Iraqi secular opposition. "The Badr Brigade [a pro-Iranian military unit recruited from exiled Iraqis] never came. We talked to the Iranians. They swear by the Koran that they didn't send the pictures."

This denial is echoed by an Islamic Iraqi exile in Iran itself, who exclaims bitterly at the behavior of the Iranians in 1991. "They encouraged the uprising and then betrayed it. They only let a few people cross the border to help, and they would not let them bring arms. They certainly did not put up posters—they were terrified of the American reaction." Whether or not Saddam put up fake posters, he certainly made a crafty effort to publicize the pro-Iranian element in the uprising by releasing eleven members of the al-Hakim family, well known for their alliance to the Tehran regime, in the middle of the crisis. They had been imprisoned in secret for a decade and the outside world had believed them long dead.

Wiser heads among the insurgents in the south knew that everything depended on the Americans. A fatal miscalculation by the U.S. commander in chief, General Norman Schwarzkopf, had allowed the bulk of Saddam's most loyal and proficient military units, the Republican Guards, to escape an allied encirclement twenty-four hours before George Bush called his cease-fire. Unlike the bulk of the Iraqi army, the Republican Guards were not conscript cannon fodder. Schwarzkopf's failure to intercept the guards was to have profound consequences for Iraq. Carefully recruited, well paid, and lavishly equipped, most of them had stayed together while the rest of the army in the south disintegrated. They would be a formidable force against the enthusiastic but chaotic insurgency that had seized control of southern Iraq.

"The biggest reason for the intifada is that they [the rebels] thought the Americans would support them," says Sayid Majid. "They knew they couldn't beat Saddam on their own. They thought they could get control of the cities and that the Americans would stop the army from intervening."

On March 9, Hussein Kamel, Saddam's cousin and son-in-law, began the counterattack on Kerbala, the other great city of Shia Islam, sixty miles from Najaf. He used Republican Guard units that had escaped the allied offensive almost intact. Brigadier Ali and other rebel officers went there to help the resistance, but as the Republican Guard tightened its grip around the holy city and terrified civilians fled to nearby villages, he realized that it was the beginning of the end. On the roads out of the city, Iraqi army helicopter crews poured kerosene on the columns of fleeing refugees and then set it alight with tracer fire. American aircraft circled high overhead.

"We had the message that the Americans would support us," lamented the brigadier as he relived his escape back to Najaf from Kerbala in a quiet North London office seven years later. "But I saw with my own eyes the American planes flying over the helicopters. We were expecting them to help; now we could see them witnessing our demise between Najaf and Kerbala. They were taking pictures and they knew exactly what was happening."

Back in Najaf, which itself was about to come under attack, Ali and the other officers consulted with Ayatollah al-Khoie. The venerable Shiite religious leader endorsed the notion that they should go south and contact the allies. "Find out what were their ideas about us, what were they going to do?" He agreed that Sayid Majid should go with them.

As Ali and the little group drove through the towns and villages of anarchic southern Iraq, their car was besieged by crowds who had heard that a son of al-Khoie had come among them. People clamored for arms. The Americans, they said, had stopped the rebels in the river town of Nassariyah taking desperately needed guns and ammunition from the army barracks. In other places, U.S. army units were blowing up captured weapons stores or taking them away. Above all, the rebels wanted communications equipment. Although they had captured almost all of southern Iraq, the successful rebels in individual cities were barely in touch with each other.

Outside Nassariyah, they met their first Americans. They were soldiers manning M-1 Abrams tanks and Bradley armored personnel carriers, part of the huge force that had swept around Kuwait and deep into Iraq in the lightning allied offensive at the end of February. The Iraqis explained to the American commander who they were and why they were there. It was not a warm reception. The U.S. officer went away for ten minutes and then returned with the curious claim that he was out of touch with his headquarters. To a professional military officer like Ali, this seemed highly unlikely. The American curtly suggested that they try and find the French forces, eighty miles to the west.

Bitterly disappointed at this disinterest, the Iraqis went in search of the French. When they eventually found them on March 11, their luck appeared to change. The lieutenant colonel in charge questioned them in detail through an Algerian interpreter and then said he would get in touch with the allied command. He seemed well aware of the significance of the al-Khoie name. Four hours later, he came to report that General Schwarzkopf himself would meet them at Safwan in two days. Safwan was two hundred miles away, too dangerous for a drive across country strewn with Iraqi government units. Could they not use one of the helicopters they could see constantly taking off from the base? asked the little delegation. At first the French told them that a helicopter would be available. For three nights they waited with growing frustration at Samawa, continually being told that the meeting with Schwarzkopf would be delayed. Majid recalls that in conversation the French told them: "The Americans are worried about the Iranians. They asked who brought Khomeini's pictures into Iraq. I explained that I had seen no pictures of Khomeini in any of the cities I had passed through. I said that people were mistaking pictures of my father, Grand Ayatollah al-Khoie, for Khomeini because both were old men with white beards and turbans."

Finally the answer came from the Americans. "We were told they had canceled the meeting in Safwan and that they would not send a helicopter." Majid knew then that the revolt was doomed.

Saddam knew it already. Twelve miles or so north of Baghdad, a heavily guarded compound at al-Rashedia houses the headquarters of the signals intelligence agency that monitored all electronic communications, including calls on satellite phones. Sometime during the first week of the intifada, military intelligence chief Wafiq al-Samarrai was handed a transcript of two radio conversations in southern Iraq that had just been intercepted by one of the al-Rashedia listening posts. Following the procedure for especially urgent intelligence, a copy had already been rushed directly to Saddam.

The intercepted conversations were between two Islamic rebels somewhere near Nassariyah. As recalled by al-Samarrai, they went as follows: "We went to ask the Americans for their support," reported one of the speakers. "They told us, 'We are not going to support you because you are from the al-Sayed group [that would be Mohammed al-Hakim].'"

"Ask them again, go back and ask once more."

The reply soon came. "They say, 'We are not going to support you because you are Shia and are collaborating with Iran.'"

The American terror over Iranian intervention had condemned the uprising. If the Iraqi leader had indeed organized the distribution of Khomeini's portrait in the insurgent towns, his ruse had succeeded brilliantly. In any event, it was a turning point. Saddam knew he might be saved after all.

"After this message," says al-Samarrai, "the position of the regime immediately became more confident. Now [Saddam] began to attack the intifada."

The first city to fall was Basra, after a mere week outside Saddam's control. In the flatlands of the Mesopotamian plain, which stretches five hundred miles, from the Kurdish mountains to the Gulf, the mechanized forces of the Iraqi army could always outflank and surround the rebels. Iraqi tanks from the Fifty-first Mechanized Division, one of the few units apart from the Republican Guard that escaped mutiny after the Kuwait debacle, quickly captured the main road overlooking the sprawling working-class slums of North Basra. The low brick houses provided little protection against heavy machine gun bullets. The tanks fired shells into centers of resistance like the local fire station, which burned to the ground. "I would say there were more than one thousand dead," said Dr. al-Rawi six weeks later. "Basra General Hospital issued six hundred death certificates. It was a bad time. You could see dogs eating bodies in the streets."

On the mid-Euphrates plain, the fighting was even fiercer. Republican Guard tanks led by Hussein Kamel were first held at al-Aoun, east of Kerbala, but they circled behind the city, cutting it off from the south. To deny the rebels cover for ambushes, the army chopped down or burned palm groves beside the roads. By March 12, says one of those fighting inside the city, "Kerbala was finished, although resistance went on until March sixteenth." Artillery and tank guns systematically blew up the shops and small workshops between the shrines of al-Abbas and al-Hussein, which stand four hundred yards apart. A rocket-propelled grenade hit the blue-and-yellow tiles of the outer porch of the shrine of al-Abbas, the warrior-martyr of Shia Islam. One memorial of the uprising carefully preserved by the Iraqi troops was a rest room for pilgrims in the al-Abbas shrine, where a noose hung from the ceiling. Here, government officials later explained, pointing to bloodstains on the floor, the rebels hanged or hacked to death Baath Party members.

In every city they captured, the soldiers immediately posted pictures of Saddam Hussein. At the shrine of Imam Ali in Najaf, where mortar fire had chipped the stones in its courtyard, soldiers placed a strangely inappropriate picture of the Iraqi leader on a chair beside the rubble. He is portrayed dressed in tweeds, walking up an alpine slope, a scene reminiscent of *The Sound of Music*. He smiles as he leans down to pick a mountain flower.

In recaptured cities across southern Iraq, government forces exacted immediate revenge. Grand Ayatollah al-Khoie and his son Mohammed Taqi were taken to Baghdad where, after a night spent in the military intelligence headquarters, they were summoned to a two-hour meeting with an angry Saddam Hussein. As later recalled by Mohammed, who sat silent and let his father do the talking, Saddam said: "I didn't think you would do something like this." The old man replied that he had wanted to control the violence. Saddam replied: "No, you wanted to overthrow me. Now you have lost everything. You did everything the Americans wanted you to do."

While Grand Ayatollah al-Khoie was in Baghdad, some 102 of his students and followers disappeared, never to be seen again. He himself was sent back to a heavily guarded house in Kufa where he remained, lying on a divan, under effective house arrest. Presented by the government to foreign visitors, he would only say ambiguously: "What happened in Najaf and other cities is not allowed and is against God." He told us, "Nobody visits me, so I don't know what is happening. I have trouble with my breathing."

Punishment for involvement in the uprising often took the form of a bullet in the back of the head. But as striking as the atrocities was the casual violence with which Baath Party leaders disposed of suspected enemies. The party always has had a cult of toughness and political machismo, which makes it sometimes record its more violent actions on film to encourage supporters and frighten its enemies.

In March 1991, the party took just such a film of Ali Hassan al-Majid, newly appointed interior minister, cousin of Saddam and known as "Ali Chemical" in Kurdistan for his use of gas against the Kurds in 1988. It showed him and other party leaders hunting down rebels in the flat and marshy lands around the town of Rumaytha, between the cities of Najaf and Nassariyah in the south.

In the film, al-Majid, who was also briefly governor of Kuwait, shows as little mercy to the Shia as he did to the Kurds. He tells an Iraqi helicopter pilot on his way to attack rebels holding a bridge: "Don't come back until you are able to tell me you have burnt them; and if you haven't burnt them, don't come back." At one moment he is joined by Mohammed Hamza al-Zubeidi, who later became prime minister, his reputation enhanced because of his toughness during the uprising. He kicks and slaps prisoners as they lie on the ground, saying: "Let's execute one so the others will confess." The prisoners, all in civilian clothes, look frightened and resigned. They are silent, except to say softly: "Please don't do this." There is the crackle of machine-gun fire in the background. Al-Majid, who looks a little like Saddam Hussein, chain-smokes as he interrogates prisoners. Of one man he says: "Don't execute this one. He will be useful to us." The soldiers, from an elite unit, shout "Pimp" and "Son of a whore" at another prisoner.

By March 16, Saddam felt confident enough to address his people in a broadcast speech. He explained that he had said nothing immediately after the war, preferring to wait "until tempers had cooled. . . . In addition, recent painful events in the country have kept me from talking to you." He blamed the southern uprising on Iranian agents—

"herds of rancorous traitors, infiltrated from inside and outside the country"—while reminding his Kurdish listeners that "every Kurdish movement linked to the foreigner . . . brought only loss and destruction to our Kurdish people," and that neighboring countries would never permit independence for Iraqi Kurds out of fear of their own Kurdish populations, a valid point. He portrayed himself as the one obstacle to Iraq turning into another Lebanon and endangering the ruling Sunni minority.

In any case, the Kurds had far outrun what they could defend. Massoud Barzani and Jalal Talabani had some fifteen thousand "Peshmerga" (Kurdish guerrillas) when they started their offensive. They were joined by over a hundred thousand Jash militia belonging to tribes allied to Saddam, as well as the many Kurds who had deserted from the Iraqi army. But the Jash turned against Saddam primarily because they thought he was going to lose. A few weeks later, as Saddam redirected toward the north the forces that had crushed the south, this looked less certain. The allies were withdrawing, Saddam retained his grip on Baghdad, and he had retaken Basra and Kerbala. The capture of Kirkuk, where the first Iraqi oil field began production in 1927, helped galvanize the Sunni core of the regime. They were not prepared to cede control of this vital oil region to the Kurds. A few months later, Izzat Ibrahim al-Dhouri, the Iraqi vice president, admitted to a Kurdish delegation that "only when you Kurds took Kirkuk was it possible to mobilize against you." The Kurds were faced with an insoluble military problem: Two of the largest cities they had captured, Kirkuk and Arbil, sit on the plain below the mountains. They are indefensible by guerrillas armed with light weapons to fight tanks backed by artillery. Sulaimaniya and Dohuk, the other Kurdish provincial capitals, are almost equally vulnerable. But these cities are home to most of the 3 million Kurds. The Peshmerga, even reinforced by the Jash, could not retreat into the mountains and deep gorges of Kurdistan and abandon their families in the cities. Instead, they had to flee together. Massoud Barzani recalled later that just before the Iraqi counterattack on March 29 he reviewed thousands of Kurdish volunteers near Rawanduz, in the heart of Kurdistan. A few days later, all had disappeared. He was reduced to defending a vital pass near his headquarters at Salahudin with his own bodyguard. For years, a burned-out Iraqi tank marked the spot where they stopped an Iraqi armored column.

Iraqi helicopters threw flour on the retreating Kurds, giving the impression that they were using chemical weapons. The object was to induce panic on a population with bitter memories of Saddam's lavish use of chemicals on them only three years before; it succeeded all too well. A million Kurds fled into Iran and Turkey.

By the end of March, Saddam Hussein had retaken all of the south. Samarra, the last town under the control of the rebels, fell on March 29. The Republican Guards entered Sulaimaniya, the last city held by the Kurds in the north, on April 2. Saddam had survived the great rebellions, though only by a whisker. He was so short of equipment that the tanks that finally retook Kerbala were old British Chieftains captured from Iran. Ammunition almost ran out. "We had lost two hundred and five million bullets in Kuwait. When we asked the Jordanians for a few million, they refused," recalls Wafiq al-Samarrai. "By the last week of the intifada, the army was down to two hundred and seventy thousand Kalashnikov bullets." That was enough for two days' fighting.

It had been a narrow escape, but now that Saddam had surmounted the immediate threats of the allies and the uprisings, something of the mood of messianic self-confidence with which he had invaded Kuwait eight months before returned. "Things are not so bad," he said to a confidant after the tide had turned. "In the past, our enemies have taken advantage of our mistakes. In future, we will sit back and take advantage of the mistakes made by them."

Saddam seemed to believe that he could now return to something like the status quo of August 1, 1990, the day before the invasion of Kuwait. But his world had changed. The United States and its allies, principally Great Britain, were determined at the very least that Saddam, their erstwhile ally, should never again be in a position to threaten their interests in the Middle East. Prior to August 1990, he had been left to deal as he wished with his own people even as multibillion-dollar oil sales financed his grandiose ambitions. At the end of March 1991, even with the rebellions suppressed, his domain had shriveled. The economic sanctions forbidding the country's vital oil exports as well as all other normal commerce with Iraq had been imposed by the United Nations Security Council on August 6, 1990. Their original purpose had been to force Iraq out of Kuwait. But, even now that Kuwait had been liberated by the allied armies, the sanctions were still in place. If they were not lifted, Saddam's income—and the standard of living of ordinary Iraqis—would be decided at United Nations headquarters in New York and Washington. Iraq would no longer be a fully independent state. For Saddam Hussein to survive under these circumstances, his enemies would have to make a lot of mistakes.

TWO

"We Have Saddam Hussein Still Here"

Three months to the day after the allied guns fell silent in Kuwait, a highly classified letter landed on the desk of Frank Anderson, a gray-haired senior official at CIA headquarters in Langley, Virginia. Anderson looked at it glumly and then scribbled "I don't like this" in the margin.

The letter was a formal "finding," signed by President Bush, authorizing the CIA to mount a covert operation to "create the conditions for the removal of Saddam Hussein from power." Anderson, as chief of the Near East division of the agency's Directorate of Operations, was the man who would have to carry it out. He was being asked to succeed where seven hundred thousand allied soldiers had failed and he did not think it could be done. "We didn't have a single mechanism or combination of mechanisms with which I could create a plan to get rid of Saddam at that time," he said later.

CIA officials faced with peremptory orders to deal with some foreign irritant—as in "Get rid of Khomeini"—like to quote an aphorism coined by a former director, Richard Helms: "Covert action is frequently a substitute for a policy." Anderson was paying the price for the war planners' failure to think about the future of Iraq after an allied victory in Kuwait.

George Bush himself had been the first to express the notion that the war might have been a triumph without a victory. "To be very honest with you, I haven't yet felt this wonderfully euphoric feeling that many of the American people feel," he said the day after his armies ceased fire. "I think it's that I want to see an end. And now we have Saddam Hussein still here."

Bush had ordered the cease-fire because his armies had overrun Kuwait in a headline-friendly 100 hours with minimal casualties. It appeared to have been the military equivalent of a perfect game in baseball and the American generals were not anxious to mar the record with any further fighting. In any event, the White House had been assured that the Republican Guard, Saddam's most loyal and accomplished troops, were trapped without the possibility of escape— one of the principal wartime objectives of the U.S. military command.

In fact, even before Bush called a halt, the bulk of the Republican Guard had already eluded the planned allied encirclement with relative ease, moving out of the intended area of entrapment on February 27. By March 1, they were sixty miles north of Basra, therefore a delay of twenty-four hours in announcing the cease-fire would have made no difference. It was only one of many miscalculations by the U.S. war planners. Other objectives wrongly thought to have been achieved included the severing of Saddam's communication links with his troops and the destruction of Iraq's nuclear, biological, and chemical warfare programs. "Saddam Hussein is out of the nuclear business," Defense Secretary Richard Cheney had confidently asserted to a closed hearing of the Senate Foreign Relations Committee after weeks of bombing. Like many other assumptions about the consequences of the Iraq campaign, this boast was soon to be revealed as embarrassingly false.

Years later, Bush would still be haunted by the recurring question: Why had he not "gone all the way to Baghdad" and settled the Saddam problem when he had had the chance? Each time he would patiently explain that the United Nations resolutions under which he had launched the war authorized only the liberation of Kuwait and he could not legally have gone further. Iraqi resistance would have stiffened. And anyway, if the Americans had gotten to Baghdad, they would have had to occupy the place for months afterward.

That was not quite the whole story. As British diplomats from the Gulf had forcefully pointed out in a secret meeting before the war, if the allies displaced Saddam and occupied Baghdad, they would eventually have to hold elections for a new government before pulling out. This would have led to all sorts of problems for Anglo-American allies among the semifeudal monarchies of the region, especially Saudi Arabia. No one wanted to encourage democracy in Iraq. It might prove catching. It had been a conservative war to keep the Middle East as it was, not to introduce change.

Militarily, an advance on Baghdad might not have been difficult. General Steven Arnold, the U.S. Army's chief operations officer in Saudi Arabia, actually drew up a secret plan after the cease-fire entitled "The Road to Baghdad," which he calculated could easily be carried out with a fraction of the forces available. Arnold's commanding officer, horrified at such an implicit admission that the victory was less than complete, put the plan under lock and key. Unfortunately, neither the military nor the White House had as yet any other plan for dealing with Iraq once the issue of Kuwait had been settled.

According to Chas. Freeman, wartime ambassador to Saudi Arabia, this lack of forethought was deliberate. "The White House was terrified of leaks about any U.S. plans that might unhinge the huge and unwieldy coalition that George Bush had put together to support the war," he recalled later. "So officials were discouraged from writing, talking, or even thinking about what to do next."

Faced with such awkward considerations, the conduct of the war had been left largely to the military, whose vision had its limitations. Before the bombing started, an air force general paid a call on Ambassador James Akins, a distinguished former diplomat with a wealth of experience in Iraq. The general explained that he wished to consult the ambassador on the selection of suitable bombing targets. Akins suggested that the Pentagon might find it more useful to draw on his knowledge of Iraqi politics and of Saddam, whom he had known for years. "Oh, no, Mr. Ambassador," said his visitor. "You see, this war has no political overtones."

During the war itself, the U.S. high command pursued a straight-forward approach to Iraqi politics: Kill the president of Iraq. The chosen weapons were laser-guided bombs aimed at Saddam's command posts, meticulously charted by the targeters. Since the United States has officially foresworn assassination as an instrument of foreign policy, the scheme was cloaked in euphemisms about targeting "command and control" centers. Nevertheless, the killing was scheduled from the day in August 1990 when air force planners wrote "Saddam" as the main priority in the first bombing plan. The air force chief of staff was fired a month later for publicly admitting that the Iraqi leader was "the focus of our efforts."

Brent Scowcroft, Bush's National Security Adviser and trusted confidant, conceded afterward that "We don't do assassinations, but yes, we targeted all the places where Saddam might have been."

"So you deliberately set out to kill him if you possibly could?" he was asked.

"Yes, that's fair enough," replied the man who had approved the hit. In fact, the Iraqi leadership, anticipating the Americans' intentions, knew full well that the most dangerous place to be during the war was inside a bunker. Most stayed in suburban houses in Baghdad. "They weren't huddled in a bunker," says a senior Iraqi officer, "because we were well aware that they were well known to the allies. We also knew that there were weapons that could destroy them."

The hunt petered out after one of the places targeters thought their quarry might be turned out to be the Amariya civilian bomb shelter and over four hundred people, mostly women and children, were incinerated. The generals' fixation on targets was unfortunate because, while U.S. intelligence knew a great deal about Iraq—buildings, communications systems, power plants, bunkers—it knew very little about Iraqis. For many years there was no U.S. embassy in Baghdad and, in any case, the country and its people were screened from the outside world by an efficient and ruthless regime. Even when Saddam Hussein needed the help of U.S. intelligence, he had done his best to keep the Americans in the dark as much as possible about events in his ruthlessly efficient police state.

In the 1980s, the two countries had been de facto allies—full diplomatic relations were restored in 1984—in the war with Iran, and the CIA sent a liaison team to Baghdad to deliver satellite photos and other useful intelligence. It was a handsome gift, but Saddam, the seasoned conspirator, was highly sensitive to the perils of such a relationship.

From 1986 on, General al-Samarrai, then deputy head of the Istikbarat, military intelligence, was one of only three officers permitted by the dictator to meet with the CIA. Just to be on the safe side, Saddam put al-Samarrai himself under intensive surveillance by the Amn al-Khass, the special security organization that reported directly to the presidential palace.

"The CIA used to send us a lot of information about Iran," al-Samarrai remembers. In addition, when preparing for an attack, his service would routinely request specific intelligence from the Americans. "I used to say, for example, 'Give us information on the Basra sector.' Saddam would say: 'Don't tell them like that, ask them to give us information from the north of Iraq to the south, because if we tell them it's only Basra, they would tell the Iranians.'" al-Samarrai would sometimes get memos on his U.S. contacts back from his master with cautionary notes scribbled in the margins: "Be careful, Americans are conspirators."

(Saddam's suspicions were not without merit. In 1986, during the infamous Iran-Contra episode, the United States gave the Iranians intelligence on the Iraqi order of battle. Coincidentally or not, Iraq then suffered a stunning defeat in the Fao peninsula.)

Late in 1989, the war with Iran won, Saddam decided that the relationship had outlived its usefulness, and expelled the CIA officials stationed in Baghdad. Diplomats who remained until the invasion of Kuwait were hardly better situated to collect information, since all contacts with ordinary Iraqis were tightly restricted. Even maids and chauffeurs catering to the diplomats' domestic needs tended to be foreign workers, Egyptians or Palestinians. In any case, all contacts with foreigners were subject to suspicious scrutiny by the Mukhabarat.

After the invasion of Kuwait, the various U.S. intelligence agencies speedily accumulated a vast quantity of information from surveillance satellites and spy planes. A massive CIA program to interview the foreign contractors who had helped build the bunkers, radar sites, communications links, and other physical infrastructure for Saddam's war machine produced further mountains of reports. Sometimes the methods used were ingenious, as when the CIA analyzed the clothes of former American hostages who had been held at the Tuwaitha nuclear plant and found telltale flecks of highly enriched uranium, a clear indication of an Iraqi bomb program.

The most secret component of the collection effort was the small group of agents recruited and infiltrated into Iraq. Given the consequences of being caught, these were courageous individuals. Communication was difficult; the radios with which they were provided did not always work efficiently, and some among the spies were reluctant even to take the risk of switching the devices on. "One or two of them were very useful," recalls one former CIA official involved in the program. On the other hand, the high command in Riyadh gave the final order to attack the Amariya shelter only after a "reliable" agent reported that it was being used for military purposes.

Astonishingly, one potentially fruitful source of intelligence was off limits. In 1988, the Iraqi and Turkish governments had complained when a mid-level State Department official received a Kurdish opposition leader to hear complaints about Saddam's use of poison gas against his subjects in Kurdistan. Any implicit recognition of Kurdish nationalism was anathema to either regime, so in deference to the sensitivities of these two allies, Secretary of State George Shultz had thereupon forbidden all further contact by any official of the U.S. government with any member of the Iraqi opposition.

The "no contacts" rule still applied during the war, which was why, for example, an offer of timely military intelligence from the Kurdish underground in northern Iraq was spurned by the Pentagon. Eventually, a system was improvised by which reports collected by Kurds were radioed to their office in Iran, thence to Damascus, thence by phone to another office in Detroit, and then faxed to Peter Galbraith, the sympathetic staff director of the Senate Foreign Relations Committee. "This was not stupid stuff," remembers Galbraith. "One of them was about what happened to an allied pilot who had been shot down. But they were picked up by a bored lieutenant from naval intelligence who couldn't have been less interested."

On the day that the allied forces ceased fire, February 28, 1991, the Kurdish leader Jalal Talabani tried to enter the State Department, intending to brief officials on the imminent uprising in northern Iraq. Thanks to the bar on contacts, no official dared speak with him, and he and his party never got beyond the department's lobby. The following day, Richard Haass, director for Middle East Affairs on the National Security Council staff, phoned Galbraith to complain about the Senate staffer's sponsorship of the unwelcome Kurds. Surely, protested Galbraith, the Kurds were allies in the fight against the Baghdad regime. "You don't understand," fumed the powerful White House official. "Our policy is to get rid of Saddam, not his regime."

The word "policy" was misused. In lieu of intelligence about the political situation in Iraq, the White House was acting on the basis of assumptions. Principal among these was a deeply ingrained belief that Saddam would inevitably be displaced by a military coup. A veteran of CIA operations in Iraq explains it this way: "All the analysts in State, CIA, DIA, NSA were in agreement with the verdict that Saddam was going to fall. There wasn't a single dissenting voice. The only trouble was, they had no hard data at all. Their whole way of thinking really was conditioned on a Western way of looking at things: A leader such as Saddam who had been defeated and humiliated would have to leave office. Just that. Plus," sighs the former covert operator, "none of these analysts had ever set foot inside Iraq. Not one."

"A collective mistake," agrees one former very high-ranking CIA official. "Everybody believed that he was going to fall. Everybody was wrong."

Nothing illustrates the lack of understanding of the situation on the ground in Iraq better than the notorious call by Bush that helped incite the uprising. According to sources familiar with the background of the speech, the original intent had been to send a message of encouragement to any potential coup plotters in Baghdad. Accordingly, Richard Haass drafted a call for the Iraqi military to "take matters into their own hands" and force Saddam from power. The appeal was due to be delivered by the president in the course of a speech on February 15.

Early on the morning of the appointed day, Saddam gave the first hint that he might be prepared to withdraw from Kuwait. Network news pictures of Iraqis enthusiastically celebrating the possibility of peace by firing guns in the air made a considerable impression on the

White House. It seemed there was a public opinion in Iraq after all.
A few extra words were added to Bush's script. Speaking to the American Association for the Advancement of Science later that morning,
Bush now referred to the "celebratory atmosphere in Baghdad"
reflecting the Iraqi people's desire to see the war end. Then he moved
on to appeal to "the Iraqi military and the Iraqi people to take matters
into their own hands—to force Saddam Hussein the dictator to step
aside . . . and rejoin the family of nations." Just to make sure the message got across, Bush repeated it, word for word, in a second address
that day at the Raytheon missile plant in Massachusetts.

As intended, the call for revolt got wide play on the international
news channels avidly consumed by Iraqis. The audience, however,
missed the nuanced references to "the Iraqi military *and* the Iraqi
people." They took the American leader's words at face value, drawing the reasonable conclusion that they were being called upon to
join the fight against Saddam.

The supreme irony is that Bush and his advisers, in trying to pro-
mote a coup, instead encouraged an uprising that may have prevented the very coup they so devoutly desired. An Iraqi source,
privy to the highest levels of the military at that time, has assured us
that there was indeed a coup being planned by senior generals from
some time during the war and after. But the plotters were deterred
from taking action by the Shia uprising. As members of the ruling
Sunni minority, they feared the consequences of Shia success and
thought it more expedient, for the time being, to rally around Saddam. What their attitude might have been had the United States
signaled support for the rebellion is not recorded.

George Bush himself later sensed part of the truth. In 1994, he
wrote that "I did have a strong feeling that the Iraqi military, having
been led to such a crushing defeat by Saddam, would rise up and rid
themselves of him. We were concerned that the uprisings would
sidetrack the overthrow of Saddam by causing the Iraqi military to
rally around him to prevent the breakup of the country. That may
have been what actually happened."

There is, however, another irony that Bush evidently fails to appreciate. In that first crucial week of March 1991, Saddam's fate hung in
the balance. Many ranking military commanders as well as other offi-

cials in the regime were contemplating abandoning the sinking ship and throwing in their lot with the rebels. But this was still a highly risky gamble, since the consequences of picking the losing side would inevitably be terminally unpleasant. For anyone making the choice, the attitude of the Americans was crucial. To tip the balance, Bush did not have to launch his armies on the road to Baghdad; a hint of support or even encouragement to the rebels would probably have been enough. Instead, Washington and the U.S. military command in Riyadh not only gave indications, such as allowing Saddam's helicopters to fly, that they were less than interested in the rebels' success, but also explicitly told rebel emissaries that there would be no support—as Saddam quickly discovered. In Baghdad and elsewhere, the waverers drew the appropriate conclusions.

This adamant repudiation of the rebel cause was based on another iron-clad assumption on the part of the Washington policy makers: a deeply ingrained belief that civil disorder would inevitably sunder Iraq. Since before the war, classified memos had hurtled around the national security bureaucracy, replete with ominous warnings of the consequences that would follow an Iraqi breakup, up to and including, as one Pentagon missive suggested, "the Iranian occupation of any part of Iraqi territory . . . Iraqi disintegration will improve prospects for Iranian domination of the Gulf and remove a restraint on Syria." Reports that portraits of Ayatollah Khomeini were being put up in liberated areas did not help matters. No one was going to assist what appeared to be surrogates for the dreaded fundamentalist Iranians.

As a result, the U.S. forces in the large portion of Iraq occupied during the ground offensive at the end of the war not only made no move to assist the insurgents, they actually gave tacit assistance to Saddam's forces by preventing rebels from taking desperately needed arms and ammunition in abandoned Iraqi stores. Much of these captured stocks were destroyed, but, paradoxically, the CIA took possession of an appreciable quantity and shipped it off to fundamentalists in Afghanistan, favored agency clients in the civil war in that country.

Since the president had publicly encouraged the uprising on which they were now turning their backs, the White House was embarrassed enough to draft their Saudi allies as an alibi. The Saudis, murmured officials in background briefings, were adamantly opposed to aiding the Shia, since they were in such mortal terror of Iran. Bush himself may even have believed this explanation. "It was never our goal to break up Iraq," he wrote later. "Indeed, we did not want this to happen, and most of our coalition partners (especially the Arabs) felt even stronger on the issue."

This was not, in fact, the attitude of the Saudis at the time. "The idea that the Saudi tail was wagging our dog is just bullshit," says one official who visited Riyadh in mid-March. He had been closely cross-questioned by Prince Turki bin Feisel, head of Saudi intelligence, about ways to aid the opposition (about whom the prince was woefully ignorant).

"The behavior of the Iraqi Shia in the Iran-Iraq war convinced the Saudis that the Shia were not Iranian surrogates," says Ambassador Freeman. "Washington was obsessed by that idea, and attributed it to the Saudis. I don't know where all this panic about the breakup of Iraq came from. After all, Mesopotamia has been there for quite a while— about six thousand years. Iraq is not a flimsy construction."

On March 26, 1991, Bush convened a meeting of his most senior advisers at the White House to make a final decision on help for the rebels. There was no public pressure to do so—the country was in "yellow ribbon mode," as one official remarked—and Bush himself had now joined in the general euphoria. A few days earlier, at the Grid-iron Club's chummy annual get-together of politicians and media, the "agony" of the president's wartime experience had been compared to that of Abraham Lincoln by a fawning member of the press.

At the White House meeting, a hard-and-fast decision to leave Iraq to its own devices was approved by all. Of those present, only Vice President Dan Quayle showed the slightest concern about allowing Saddam Hussein to go on slaughtering the insurgents without hindrance. No one appears to have challenged the presumption that a rebel victory would inevitably have led to Iran seizing a piece of Iraq.

Following the meeting, as Bush's spokesman announced that "We don't intend to involve ourselves in the internal conflict in Iraq," Brent Scowcroft and Richard Haass boarded a plane for Riyadh to spread the word in the field. The Saudis were still in a mood to help the rebels—a senior Kurdish representative was in Riyadh when the

Americans landed. They needed to be told to get in step with policy.

In Washington, a "senior official" was briefing reporters on the fact that Bush believed "Saddam will crush the rebellions and, after the dust settles, the Baath military establishment and other elites will blame him for not only the death and destruction from the war but the death and destruction from putting down the rebellion. They will emerge then and install a new leadership." That was not quite the picture of White House policy the Saudis got from their high-powered visitors, as Sayid Majid al-Khoie soon discovered.

Al-Khoie had been held under comfortable house arrest at the Saudi border ever since he had escaped from Iraq, the promise of his meeting with Schwarzkopf still unmet. He was there when George Bush was asked, on the day after the crucial March 26 meeting, if any rebel groups had asked the United States for help.

"Not that I know of," the president blithely replied. "No, I don't believe that they have. If they have, it hasn't come to me."

After finally being allowed to travel to Riyadh, al-Khoie had his first chance to meet with the Saudi intelligence chief Prince Turki bin Feisel on March 30, three days after the two emissaries had arrived from Washington.

Al-Khoie recorded the two-hour meeting in his diary: "Why are you so worried about the Shia?" he asked.

"We can't do anything to help you," replied the prince. "The Americans don't want to remove Saddam. They say, 'Saddam is under control. This is better than somebody we don't know about. We are worried about Iran.'"

Twenty-four hours after al-Khoie heard that the Americans now wanted Saddam to stay in power, Peter Galbraith was fleeing for his life from the Kurdish city of Dohuk. The energetic Senate aide had been touring the war-torn region and had gone to bed late the night before after telling a crowd of Kurdish notables that, as the first representative of the U.S. government in a free Kurdistan, he was proud to address them. Now he was running from a vengeful Iraqi army on the verge of retaking the city. An angry red-haired Peshmerga stuck his head through the car window. "Damn Bush," he said.

The 2 million Kurds who joined Galbraith in flight were about to upset the White House's determined disengagement from Iraqi affairs. The Shia in the south had fled in equal terror, but without attracting much attention or sympathy in the outside world. The Kurds fared better, being easily accessible to the media army that speedily materialized on the Turkish border and telegenic besides. "They look middle class," murmured a Senate staffer watching TV pictures of doctors and lawyers in three-piece suits shivering on the bleak mountain sides. "I never realized they were like us." Influential figures such as the columnist William Safire, a champion of their cause since the days of their betrayal by the CIA in the 1970s, weighed in on their behalf. Galbraith, safely over the border, threw in his own bitter and well-informed denunciations of the whole postwar policy on Iraq.

With unseemly reluctance, the White House bent to public opinion and began to assist the Kurds. At first Bush sent food and medicine and then, on April 16, he ordered U.S. troops into northern Iraq to create a "safe haven" from Saddam's forces for returning refugees.

It was a momentous turning point. Although Bush stressed that the troop deployment was merely temporary, the president had now, however unwillingly, accepted a military role for the United States inside the borders of Iraq itself. Bending to force majeure, Saddam made no effort to resist. Although the allied troops were withdrawn within three months, the Iraqi army did not permanently reassert the government's control over Kurdistan. U.S. warplanes based at Incirlik, just across the Turkish border, were now assigned to "Operation Provide Comfort"—protective air cover for the Kurds and a tangible deterrent to any effort by Saddam to crush these rebellious subjects once more.

Announcing the April decision to dispatch troops into Kurdistan, the president was defensive about his famous call to the Iraqi people, now coming back to haunt him. "Do I think that the United States should bear guilt because of suggesting that the Iraqi people take matters into their own hands, with the implication being given by some that the United States would be there to support them militarily?" he replied to one aggressive questioner. "That was not true. We never implied that." Displaying a certain economy with the truth, he went on to insist that the wartime objectives had "never included the demise and destruction of Saddam personally."

In the argument that day over the gap between presidential

rhetoric and realpolitik, no one paid much attention to Bush's casual remark that not only would there be no normal relations with Iraq until "Saddam Hussein is out of there," but that "we will continue the economic sanctions." It was the single most important statement of the day.

The UN resolutions authorizing the war in Iraq had served as a useful justification in not carrying on the war after the Iraqis were evicted from Kuwait. But the economic sanctions on Iraq had been tied by the Security Council to specific ends: an unconditional withdrawal from Kuwait, compensation for damage there, and the total elimination of all weapons of mass destruction and the facilities for making them. Following the passage of the cease-fire resolution that officially ended the Gulf War on April 3, 1991, U.S. ambassador Thomas Pickering stated explicitly that, "Upon implementation of the provisions dealing with weapons of mass destruction and the compensation regime, the sanctions will be lifted." Now the president was offhandedly rewriting the UN resolution. Saddam was on notice that it did not matter whether he observed the existing resolutions or not, he would never be allowed to export his oil freely until the day he died. In putting a once wealthy country under permanent blockade, Bush was deploying a weapon far more deadly than the TV-friendly smart bombs of the war.

This time Saddam was not the direct target, a fact spelled out by deputy national security adviser Robert Gates when he gave the sanctions decision a more formal unveiling on May 7. "Saddam is discredited and cannot be redeemed. His leadership will never be accepted by the world community. Therefore," declared Gates, "Iraqis will pay the price while he remains in power. All possible sanctions will be maintained until he is gone. . . . Any easing of sanctions will be considered only when there is a new government."

Iraq was heading into the hundred-degree temperatures of summer. The hospitals were beginning to fill up with typhoid cases and the doctors were running out of drugs with which to treat them. In Basra, heavily bombed in the war and fought over again in the intifada, children splashed in pools of sewage because the sewage pumps were broken and, thanks to the sanctions, no spare parts could be imported. So long as sanctions were in force, Iraq would remain in the third-world misery to which it had suddenly been thrust.

Sanctions had the merit, in U.S. eyes, of "containing" Saddam. With an unrepaired economy, he would never be able to cause the kind of trouble across his borders that had necessitated the Gulf War. As an additional bonus, keeping 3 million barrels of daily oil production off the international markets would keep a floor under world prices, thus helping the Saudis and Kuwaitis, who were pumping hard to pay for the war. "Saddam will remain an outcast," predicted Ambassador Akins, "unless and until oil goes back up to $30 a barrel. In that event, he will be reincarnated as Mother Teresa."

Meanwhile, as Gates observed, the Iraqi people would be paying the price. To the outside world, the administration plan seemed clear: If the Iraqi people would only rise and get rid of Saddam, then the blockade would end. It was, therefore, confusing for the present authors to be told at the time by a senior CIA official that a popular uprising was the "least likely" consequence of sanctions. In reality, the senior officials at the Pentagon, White House, State Department, and CIA who crafted the policy had a slightly different end result in mind. It was not the people who were expected to take action, but members of the ruling elite.

"They really believed that the sanctions policy might encourage a coup," says a former official with the CIA's covert operations arm who was much involved in Iraqi matters at the time. "You have to realize, they understood very little about the way that Saddam thought, and nothing about the 'fear factor' with those around him."

Up until now, the United States had tried to encourage a coup from the sidelines, as with Bush's wartime appeal and subsequent offhand remarks about getting "Saddam Hussein out of there." No one had a clear plan as to how the longed-for event might be achieved. In what one official describes as a "painfully frequent" series of meetings between the CIA and other interested parties at the Pentagon, State Department, and White House, officials had groped for answers. The air was thick with simplistic slogans. Some suggested giving full backing to the Kurds, now safe back at home under protective U.S. air cover following Bush's reluctant intervention in northern Iraq, to trigger a "rolling coup" that would sweep progressively southward from their mountain fastness. Others argued that rigorous sanctions would eventually prompt some public-spirited member of Saddam's family or bodyguard to do the deed—the so-called "silver bullet" solution. A variant on this notion propounded the prospects of a "palace coup" by disgruntled Republican Guards or other security units.

No one was foolish enough to believe that any single idea provided a guaranteed solution. So now Frank Anderson was being tasked to try a combination of these vague schemes in the hope that something would turn up. As the experienced covert operator fully understood, he did not have many tools with which to do the job.

The previous August, the president had signed another CIA finding on Iraq. Contrary to later reports, however, this was not a directive to overthrow Saddam. "Our mission was to convince Saddam that the holocaust was coming unless he backed down," explains one former CIA officer drafted for the task. "We were finding people who were going to see Saddam, asking them to pass the message on of what the military had in store for him." In addition, agency operatives helped spread propaganda in other countries "telling them what a bad guy he is, which was easy." The CIA's overriding priority, however, before and during the war, was to service the military campaign—interviewing foreign contractors, analyzing satellite intelligence (which led to sharp disagreements with the military on how much or how little of Saddam's arsenal had been destroyed), and infiltrating agents with "electronic gadgets" into the enemy capital.

There were indeed "clandestine" radio stations broadcasting into Iraq from Egypt and Saudi Arabia—"the Voice of Free Iraq"—but these, though monitored by the agency, were under the day-to-day control of local intelligence agencies. "They would put out a lot of information about how Saddam was about to fall, defections of senior officers, that sort of thing," one former agency officer remembers. "Eventually we had to tell the FBIS [Foreign Broadcast Information Service, which monitors and translates foreign broadcasts] to stop carrying their stuff because we'd get calls from the White House asking about some coup that the radio station said was under way."

The Saudi broadcasts were the actual handiwork of a group called the Iraqi National Accord (al-Wifaq), founded by two disaffected veterans of Saddam's ruling Baath Party who had later fled into exile. It was a group that was to figure largely, and fatally, in the agency's attempts to carry out the mission, and the story of its lead-ing lights serves as an instructive example of the kind of "mecha-nisms" to which the CIA would eventually turn.

One of the founders, Salih Omar Ali al-Tikriti, had once enjoyed a stellar career in Baghdad, from supervising public hangings to diplo-matic service as ambassador to the United Nations. He resigned that post in 1982 under the mistaken impression that recent disasters in the Iran war would cause Saddam to fall and that he, as a Sunni Baathist from Tikrit, could be a viable replacement. Subsequently reconciled in some fashion with Saddam, on August 2, 1990, Salih Omar had the highly lucrative post of heading the London office of Iraqi Freight Ser-vices Ltd., a front company for the Iraqi government. "His office por-trait of Saddam was wall size," remembers a fellow exile. On August 6, the day the United Nations announced economic sanctions—thus putting the freight company out of business—Salih Omar once again announced himself as a member of the opposition. He soon found favor with the Saudis (who were so ignorant of Iraq that their list of opposition leaders included names of men long dead) and moved to Riyadh.

Omar's partner in the Accord, Iyad Alawi, had also done yeoman service for the Baath Party, having been a student organizer in the days before the revolution and later moving to London. In Great Britain, he exercised a key function for Iraqi intelligence as head of the Iraqi Student Union in Europe. The Arab students with whom he came in contact were of considerable interest to Baghdad, since they tended to be drawn from elite circles in the Middle East. They were also of more direct value to Alawi personally, garnering him a fruitful array of connections in Saudi Arabia and elsewhere, which he then used to great effect in various business enterprises in the region. By the late 1970s, he had become rich.

However, Alawi never lost his taste for the world of intelligence and the company of intelligence officers. Soft-spoken, eloquent, and per-suasive, always ready to hint at a powerful connection or make a promise, he proved adept at telling them what they wanted to hear in language they could understand. In 1978, this mutual affection almost proved fatal. By that time, Alawi had reportedly entered into a rela-tionship with the British security services, who would naturally have been keen to acquire a willing and well-informed source in the huge and intrigue-ridden Arab student community in London. Word of this relationship reached the attentive ears of the Mukhabarat in Baghdad, who dispatched a team armed with knives and axes to Alawi's comfortable home in Kingston-upon-Thames to deal with the problem in summary fashion. Bursting into his bedroom, the assassins hacked at him as he lay beside his sleeping wife and were only prevented from finishing the job by the fortuitous appearance of his father-in-law, who happened to be staying in the house. The would-be killers ran off and the badly injured Alawi lived to make more money and pursue his connections with British intelligence and similar organizations.

By the time the war began, Alawi had scented the interest of Saudi intelligence and had joined forces with his fellow ex-Baathist, Salih Omar, in producing the Voice of Free Iraq. The pair soon fell out, however, reportedly because of a dispute over a $40,000 check from their Saudi paymasters. Omar gradually faded from sight, although in the days of the uprising he claimed he was in close touch with senior members of the Iraqi command who were ready to mount a coup. Alawi retained control of the Accord, into which he steadily recruited former Baathist Sunnis—the type best suited to preserving the regime post Saddam—and was soon back in London awaiting fresh clients.

The money that allegedly fostered the split between the two allies was only a tiny fraction of the sums allocated by Prince Turki bin Feisel's organization, reportedly as much as $50 million. However, Saudi Arabia had until August 2, 1990, been a staunch ally of Saddam's government and had sent its subsidies (including substantial contributions to the Iraqi nuclear weapons project) directly to Baghdad. The brigadier heading the Iraq desk at Saudi intelligence was famously ill-informed, at one point asking suspiciously, "Are the Kurds Muslim?" Far more expert in the maze of Iraqi opposition politics were the Iranians and the Syrians.

The Iranians' favored instrument was, of course, Mohammed Baqir al-Hakim and his Supreme Council (sometimes translated as Assembly) for the Islamic Revolution in Iraq, complete with its military arm, the Badr Brigade. The Iranians, no less than the Americans, were brooding about the situation following their refusal to help tip the balance during the uprising, but there was little or no prospect, as yet, for Frank Anderson and his officers to enlist them as a "mechanism" in the mission of overthrowing Saddam. Iran, after all, was universally feared in Washington as the predator waiting to acquire "chunks" of a dismembered Iraq.

The Syrians, on the other hand, were more accessible, having both an impeccable record of vicious enmity toward the rival Baathist regime in Baghdad and credentials as a member of the Gulf War coalition. Damascus hosted its own quota of exiled dissidents, former generals and government ministers, blown across the border by unremitting crackdowns and purges since Saddam had seized total control in 1979 and immediately announced the discovery of a Syrian-backed plot against him. Even before the war, in late December 1990, the Syrians had sponsored a meeting of local and visiting opposition figures who emerged with a program that called for the overthrow of Saddam and the installation of a coalition government.

The following March, with Saddam routed from Kuwait and the intifada still blazing across Iraq, the Saudis decided it was time to have another, grander gathering of the opposition, this time in Beirut. However, "Abu Turki," as Riyadh's intelligence expert on Iraqi affairs was derisively known among Iraqi exiles, did not feel qualified to arrange such a complicated affair himself. Since the Syrians had better connections across the range of opposition groups than the Saudis, he handed $27 million to his Syrian counterpart and left him to get on with the task, insisting only that some of his favorite Iraqis, such as Salih Omar, be invited along. The Syrians, apparently concluding that this was far too large a sum to be lavished on a group of Iraqi dissidents, pocketed most of the money and handed over the residue, in Syrian currency, to the Iraqis to pay for airfare and hotels.

It was the largest gathering in the history of the Iraqi opposition. All shades of opinion were there, from the ex-Baathists, to the Tehran-based Islamists, to the remnants of the once-powerful Iraqi Communist Party, as well as the Kurds. They voted by a narrow margin to escape from Syrian control and seek support in the West. The Kurdish delegation, who had good reason to be suspicious of Western support after their abandonment by the CIA in 1975, supplied the swing vote.

For many of the delegates, these were heady times. As Laith Kubba, a civil engineer who had been living in England since he left his homeland in 1976, recalls, "Following the Gulf War, the whole world wanted to know the Iraqi opposition." Kubba remembered well how different it had been for most of his exile. "When Saddam was being supplied by the United States and Great Britain, we suffered. There was a thriving industry forging extensions on Iraqi passports so that people could avoid being deported back to Iraq. Sometimes we had little victories, like the time we demonstrated against one of Saddam's ministers who was in London signing a trade deal with the Thatcher government and we forced him to leave by the back door." On one occasion, at a conference on Iraq at a London think tank, Kubba bumped into the man in charge of Middle East affairs at the U.S. State Department, a notably fatuous official named John Kelly. Kubba introduced himself as a member of the Iraqi opposition. "How long have you been working for the government of Iran?" said Kelly, before turning his back.

As we have seen, the March 1988 gassing of five thousand Kurds in the city of Halabja in a single afternoon was greeted by a thunderous silence from Western governments. Kubba took a leave from his job and spent a month crisscrossing the United States. Over and over he screened a video of the effects of the attack to anyone who was interested, a lonely effort to show the world what was going on in Iraq.

Unlike seasoned operators Salih Omar and Iyad Alawi, Kubba still retained a certain hopeful naïveté. When Saddam invaded Kuwait, he was on vacation in Florida. Following up on a chance introduction on a plane, he drove to Washington and secured an off-the-record interview with a mid-level State Department official (carefully avoiding the word "opposition"). Earnestly, he suggested that the United States use the invasion crisis to advance the cause of democracy in Iraq. "Who told you we want democracy in Iraq?" answered the official, flushed with bureaucratic machismo. "It would offend our friends the Saudis." Kubba was shocked.

Eight months later, the atmosphere at the State Department had apparently changed, at least a little. At last the U.S. government was ready to meet with the Iraqi opposition. On April 16, Kubba was officially received at the imposing building on C Street by David Mack, the deputy assistant secretary of state for Near Eastern Affairs. Mack read from a printed paper outlining U.S. government policy toward Iraq. Apart from ringing phrases concerning "sovereignty, integrity, democracy," Kubba recalls an explicit statement that "We are not involved in Iraqi politics. There will be no U.S. soldiers on Iraqi soil." Two hours later, President Bush appeared on television to announce that he had ordered U.S. troops into northern Iraq.

Kubba was one of three Iraqis to meet with Mack. Sitting beside him were Latif Rashid, a brother-in-law of the Kurdish leader Jalal Talabani who had been so brusquely treated on the day the war ended, and Ahmad Chalabi, a portly gentleman not previously marked as a committed member of the opposition.

Chalabi had a background very different from that of his companions. He came from an extremely wealthy Shia banking family who in the days before the 1958 leftist coup that overthrew the monarchy and murdered the king had been very much a part of the elite. Following that revolution, the family moved to Lebanon, where they continued to prosper, forming close links with the Lebanese Shia community. Abdul Karim al-Kabariti, later prime minister of Jordan, knew Chalabi as a young man in Beirut. He remembers him as a "walking encyclopedia—smart, but not wise." From Beirut, Chalabi went to MIT and thence to the University of Chicago, where he acquired a Ph.D. in mathematical knot theory, before returning to the Middle East and the family business. In 1977, he moved to Jordan and founded Petra Bank, which expanded rapidly in an ill-fated Jordanian economic boom of the early and mid-eighties. Soon, it was the third-largest bank in the country. Chalabi himself had friends in very high places in Amman, who were, however, of little use to him in August 1989, when the governor of Jordan's Central Bank suddenly moved to take over Petra Bank because of what were termed "questionable foreign exchange dealings." Allegations of fraud and embezzlement soon followed, and Chalabi left the country to go "on holiday," although his journey to Damascus in the trunk of a friend's car suggests a more urgent exit. Ultimately Chalabi would be convicted, in absentia, of embezzling at least $60 million (then and since he has strenuously denied the charges, which he describes as politically motivated) and sentenced to twenty-two years' hard labor. But by then he had other fish to fry.

Until the Petra Bank crash, Ahmad Chalabi had been barely involved in the politics of his native country beyond offering support to the relatives of certain Shia who had vanished into Saddam's dungeons. By 1991, however, that had changed and he threw himself into the burgeoning world of the Iraqi opposition with energy and initiative. Even before the May 1991 finding, he was registering on the CIA's radar screen as a man of whom they should take note.

According to Laith Kubba, the feeling was mutual. Not long after the meeting at the State Department, he recalls, Chalabi confided to him that "The Americans paid more than five hundred million to the Afghans. If there is a sound proposal, the United States is prepared to allocate substantial sums for the Iraqi opposition. We should go for that money."

Chalabi's arrival on the scene did not meet with universal approbation at Langley. An official who became involved in the Iraq operation in May 1991 recalls that Chalabi had already been recruited. The official also remembers Frank Anderson declaring, "I want all of the growth in this program to be not this guy."

There were good arguments against the selection of Chalabi as a "mechanism." He was, as the Americans noted, new to politics. Unlike others in the exile firmament, he had no network of supporters outside Iraq, let alone inside. He was a Shia, always an uncomfortable notion for the Americans, and, of course, there was the awkward fact that he was a wanted man in Jordan thanks to the Petra Bank scandal.

On the other hand, thought the spymasters, there was much to recommend him. "He had good organizational skills," recalls one. "Also, he had an Iraqi nationalist viewpoint, rather than just talking up the Shia cause."

Another of the CIA team points out that "There were advantages to the fact that he was a businessman, not a politician. As a businessman, he was used to thinking in terms of a program. He would start planning to print a newspaper today, and at the same time start figuring out how to get it into Baghdad six months from now. We liked that approach. Also, he was rich, which helped explain any money that we might be giving him."

Paradoxically, one of Chalabi's chief assets was his lack of a political following. "He had another advantage, in that he was weak," says one official. "All the others—Barzani [the Kurdish leader], the al-Khoie people, and so on—had power bases, but by virtue of that, they had powerful enemies within the other opposition groups. Chalabi was not a threat to anybody. He was acceptable as an office manager. So his weakness was a benefit—but then he was weak. Every single part of this had a cost-benefit analysis."

The CIA looked far and wide in its search for useful assets. Following his depressing conversation with Prince Turki bin Feisel in Riyadh, Sayid Majid al-Khoie had been suddenly flown to Paris under the auspices of people who described themselves as "French lawyers." There he found himself talking to various French and American government officials, though he was never quite sure exactly who they were. The Americans identified themselves as "State Department," but their chief interest was not Iraq but the U.S. hostages in Lebanon. There was some point to this, since Sheikh Mohammed Hussein Fadlallah, the hostage takers' spiritual mentor, was a disciple in religious matters of Sayid Majid's father, the revered ayatollah.

Though the Iraqi cleric was more concerned about the fate of his own family and friends in Najaf, he agreed to do what he could. Flying to Tehran, he met the Iranian leader Ayatollah Khamenei, who was scornful of his mercy mission, warning him that "The Americans won't help you whatever you do. Be careful with which group in America you are dealing. Are they the CIA?"

"I don't know, but if I can save only one person it is worth it," replied al-Khoie, and flew on to Beirut, where Fadlallah was equally cynical about his sponsors: "Saddam has destroyed all your cities [meaning Najaf and Kerbala]. The Americans just stood by and looked." Al-Khoie had been promised a meeting with Secretary of State James Baker. But when he finally arrived in Washington, Baker had left for Texas because his mother had died. Al-Khoie offered to wait, only to be told that, in any event, he would not be seeing the secretary.

In addition to his suspect credentials as a powerful religious Shia, al-Khoie was not cut out for a serious career as a CIA asset, displaying a prickly independence in such matters as paying his own hotel bills. Chalabi appeared a far better prospect.

Later events made it appear that the CIA were for a time pinning all their hopes on fostering a movement, headed by Chalabi, that would eventually take power in Baghdad. This was far from the case at any time. Looking at the hand he had been dealt, Anderson apparently concluded that it was extremely weak. His operation was only a part of a wider and somewhat amorphous scheme to box in Saddam through sanctions, the maintenance of a large U.S. force now permanently stationed in the Gulf, and military protection for the semiautonomous Kurdish zone in the north.

At the upper levels of the White House, there was still a lingering hope that the "silver bullet" from a disaffected bodyguard in Baghdad would put an end to all their problems. Short of that, CIA officers around the Middle East and beyond continually scanned the horizon for anyone who could offer a connection inside the Iraqi military or security forces in furtherance of the "palace coup" solution. In the meantime, as one former CIA official explains with a shrug, the idea was "a combination of sanctions and shaming him, humiliating him by showing that he did not control all his territory and was not secure. Create an ambience of 'coups and rumors of coups,' that sort of thing."

For some of the more experienced and honest officials in the upper tiers of the operations directorate, creating an ambience seemed about the best that could be done.

One problem the CIA officials in charge of the program did not face was money. Congressional intelligence committees, who did not share the weary cynicism of the official quoted here, heartily approved the notion of unleashing the CIA against Saddam and therefore authorized a budget of $40 million for the first year of the operation. Anderson was a seasoned enough Washington operator to know that Congress tends to measure action by expenditure. Therefore, his main priority was to be seen to be spending that money. As it happened, there was a convenient outlet ready to hand.

Ever since the days when the agency covertly sponsored Radio Liberty and Radio Free Europe to beam propaganda into Eastern Europe and the Soviet Union, propaganda had been an important instrument of operations. Starting in the 1970s with a successful operation in the Sudan, there had been a trend toward privatizing this activity by handing the contract to a suitable public relations firm. John Rendon, a veteran of political operations in the Jimmy Carter administration, had created what was widely considered an enormously successful propaganda campaign in softening up Panama before the 1990 U.S. invasion of that country. His first encounter with Iraqi affairs came via the bottomless coffers of the exiled Kuwaiti government after Saddam's invasion. Acting under a contract from the Kuwaitis, he had organized and run radio and TV broadcasts, beamed into the occupied emirate from Saudi Arabia, to give succor to the population languishing under Iraqi military occupation. Now the money tree shook for him again as CIA covert operations specialists sought suitable tools with which to harass Saddam. At least one of these specialists had admired Rendon's work in Panama and recommended him for work on the Iraq project.

In September 1991, Francis Brooke was looking for a job, having formerly worked as a liquor lobbyist in his native Atlanta as well as having run the Georgia state census. Now one of his political friends, who had once worked for Jimmy Carter, suggested he might try calling John Rendon, in Washington. "The place looked just like a political campaign headquarters," he recalls. "There was high-tech communications equipment strewn all over it." After a brief chat, Rendon made an offer: "How would you like to go to London to work on a program describing atrocities committed by the Iraqi army in Kuwait, at twenty thousand a month?"

Brooke paused. "Let me think about it," he said, "for about two seconds." Back home he checked newspaper files and came across Rendon's background in Panama and Kuwait. The young man's father, formerly a career military intelligence officer, heard his story and said that this was plainly a CIA operation. Nothing loath, Brooke decamped to London, where he found a large office full of people on similarly fat salaries. He was not impressed. "These were people who had difficulty in culturally adapting to London," he observes, "let alone the Middle East."

One of the projects under way was an "atrocity exhibition" of photographs and other memorabilia traveling around Europe to impress people with the heinousness of the Iraqi regime. Central to this undertaking was a sign-in book in which visitors could record their comments. The hope was that Iraqi exiles would thereby obligingly identify themselves for possible subsequent targeting and recruitment by the CIA. Other parts of the operation included a roomful of "twenty-year-olds," according to Brooke, sitting in Washington writing scripts for radio propaganda broadcasts at $100 a day. These were then shipped to Boston for translation into Arabic and then on to radio stations in Cairo, Jeddah (in Saudi Arabia), and Kuwait City to be beamed into Iraq.

"Yeah, I guess Rendon got a lot of the money," admits one of the officials involved, stressing that there was certainly no fraud involved. In the summer of 1991, there was an audit of the program. "No one found any fraud, but an agency accountant was philosophically opposed to this connection with a PR outfit, which meant that Rendon became the most audited private firm in history."

There was one person who impressed Brooke in the London office: Ahmad Chalabi. "He was the one person who seemed to know what he was doing, as well, of course, as knowing a great deal about Iraq and the Middle East." By the fall of 1991, Chalabi had made himself indispensable to the Rendon operation. Laith Kubba remembers being commissioned by him to go on the road, speaking about Saddam's atrocities. Ignorant of the suspicious accountant back at Langley—and indeed of the whole CIA involvement—Kubba could never understand Chalabi's fanatical insistence on receipts for everything, down to bus tickets.

To lower-level employees like Brooke, the entire plan appeared an exercise in futility. Such efforts as the "atrocity exhibition" seemed hardly likely to bring down Saddam. There was, however, another dimension to the operation. The price being paid by the Iraqi population because of the economic sanctions was beginning to attract a certain amount of international attention. The Finnish politician Martti Ahtisaari visited Baghdad in April 1991 and returned with a gloom-laden report predicting imminent mass starvation among Iraqis unless sanctions were lifted. A month later, a team from the Harvard School of Public Health toured Iraq and presented a more considered but hardly less alarming picture of the immense and growing suffering in a civilian population denied adequate food and medical supplies by the blockade.

Sanctions were at the center of U.S. policy as it had evolved in the first few months after the war. It was, therefore, imperative to maintain international public support for what casual readers of the Harvard team's findings and other reports might conclude was an indefensibly cruel policy. That was where the CIA operation, as deployed through Rendon's public relations exercise in Europe and elsewhere, came in useful. "Every two months or so there would be a report about starving Iraqi babies," explains one veteran of Rendon's propaganda campaign. "We'd be on hand to counter that. The photo exhibition of atrocities and the video that we had went around two dozen countries. It was all part of a concerted campaign to maintain pressure for sanctions."

Ahmad Chalabi, however, had no intention of confining his activities to PR in support of starving his fellow countrymen. He and Laith Kubba (the two are second cousins) and others had been contemplating the formation of a new Iraqi opposition group. Unlike the numerous others already in existence, this would be designed to encompass all the major factions among Iraqis opposed to Saddam and would have as an ultimate aim the creation of a democratic regime. It was called the Iraqi National Congress—INC—a name later claimed by some to have been selected by the CIA, an assertion indignantly denied by Chalabi. "It's a lie! I picked the name myself. Rendon was there when I did it."

While necessarily concealing his relationship with the agency, Chalabi had been pursuing an aggressive lobbying campaign for his cause in Washington. After impressing various members of Congress, he finally gained an audience in December 1991 with a skeptical Richard Haass, who heard him out and finally agreed that "You've given us a lot to think about." That was enough for Chalabi, who flew off to the Middle East to spread the word among the Arab intelligence services that the Americans were behind him.

It was to be a recurring pattern. For the Americans, Chalabi was a man who could speak for the Iraqis. For the Iraqis, Chalabi was becoming "the American broker."

In June, a horde of delegates flew to Vienna, Austria, for the founding meeting of the INC. The expenses were paid by the agency, a fact not known to most of the attendees. As one official helping to foot the bills remembers with a smile, "There wasn't a single person there who didn't believe he was paying for it all out of money they believed he had embezzled from the Petra Bank."

The following month, the leading lights of the INC were flown to Washington for a full-dress presentation, to meet with National Security Adviser Brent Scowcroft and Secretary of State James Baker. They were assured of American support for their organization and for a democratic Iraq, and returned home full of hope and vigor. Their hosts, however, took a more cynical view. The hopes and prospects of Iraqis who resisted Saddam were not rated as high by the professionals.

"We assumed," says one of the officials involved, "that we better view Chalabi a bit like people looked at someone in the 1950s running for statewide office in the South on the Republican ticket. Everyone knew he didn't have a chance. His job was to act like he had a serious chance."

Thus the stage was set for a tragic misunderstanding. The various factions represented in the Iraqi National Congress thought that they now had the unqualified backing of the United States government in displacing Saddam. The White House and the CIA simply regarded the INC as one more useful thorn to stick in Saddam's flesh, along with sanctions and whatever subterranean plot could be concocted to overthrow the dictator by means of a palace coup. In other words, the INC was only half of a two-pronged U.S. strategy. The INC received the public endorsement of the U.S. government because they were respectable democrats, suitably opposed to the Baath Party's control of Iraq. In the view of high officials like Baker and Scowcroft, there was no harm done in encouraging them. The INC brought an added bonus in that the adherence of the Kurds to this opposition group forestalled Kurdish moves toward independence, something that was always anathema to America's ally Turkey, facing its own Kurdish insurgency. In private, however, U.S. national security decision makers believed that only a revolt within Saddam's inner circle stood a chance of removing the Iraqi leader.

However halfheartedly, Washington was now inescapably involved in the political affairs of one of the most complex, divided, and violent societies anywhere. That society, not to mention the history and personality of the man at its center, merited closer attention than it had so far received from the outside world.

THREE

The Origins of Saddam Hussein

An Iraqi proverb says: "Two Iraqis, three sects." In Iraq, Islam does not unify, it divides. The Sunni Arabs living in the triangle of territory between Baghdad, Mosul, and the Syrian border are a fifth of the population but have always dominated Iraqi governments. The Shia Muslims make up over half the population and are the overwhelming majority in southern Iraq between Baghdad and Basra. In the capital, they outnumber the Sunni, though governments have tried to limit their immigration. In the north, the Kurds are a further fifth of the Iraqi population, living in the mountains along the Iranian and Turkish borders and the plains immediately below.

On the map, the Mesopotamian plain, stretching 550 miles from the mountains of Kurdistan to the Gulf, looks united. The Euphrates and Tigris, the rivers on which most Iraqi cities are built, never created one country in Iraq, as the Nile did in Egypt. Their shoals and shallows made navigation difficult. In the last century, it took a week to travel down the Tigris from Baghdad to Basra. When the British tried to move soldiers wounded in the battle to relieve the garrison besieged by the Turkish army at Kut in 1916 during World War I, it took thirteen days for the barges to reach Basra, only two hundred miles downriver. Mosul traded with Aleppo and northern Syria, and Baghdad and the Shia holy cities of Kerbala and Najaf were strongly connected to Iran, while in the far south, Basra looked toward the Gulf and India. At the turn of the century, Iraq was not a single political community or people. It was divided into a series of tribal federations and near autonomous cities, each with its own complex politics. Even within cities, religious and tribal divisions go deep. In 1915, the people of Najaf rose in rebellion against the Turks and expelled them. All the rebels were Shia, but even so, each of the four quarters of the city declared itself independent and remained so until the British arrived two years later.

The diversity of Iraqis is complicated by another factor, which predates the introduction of Islam. The Mesopotamian plain is a frontier zone. It is overlooked by the Iranian plateau and the mountains of eastern Turkey. It has no natural defenses and has always been the prey of the powerful states surrounding it. Iraq is full of battlefields. In A.D. 401, Xenophon and the ten thousand Greek mercenaries started their long march to the Black Sea after being defeated at the battle of Cunaxa, close to the Euphrates River and southwest of Baghdad. Seventy years later, Alexander the Great fought his decisive battle against the Persian Empire at Gaugamela, in the northern plains east of Mosul below the mountains of Kurdistan. People living in what became Iraq were in the front line in the struggles between Rome and the Persian rulers based on the Iranian plateau. The critical battle between the invading Arab armies, newly converted to Islam, and the Persians was fought at Qadisiyah on the lower Euphrates in B.C. 637. (During the Iran-Iraq war, Saddam Hussein was often referred to in the Iraqi press as "Qadisiyat Saddam.") Vulnerability to foreign invasion is a recurrent feature of Iraqi history.

The most important battle for present-day Iraqis was militarily insignificant, almost a massacre, and ended in total defeat for one side. But the details of this tragic skirmish are remembered every year in a majority of Iraqi houses. It started the conflict between the

Shia and the Sunni Muslims that divides Iraq, as it does the Islamic world, fourteen hundred years later. What began as a bloody dynastic struggle to succeed the Prophet Mohammed ended by creating rival systems of belief. In A.D. 656, a civil war erupted over who should be the fourth caliph in the newly created Islamic world, just established by a series of explosive conquests. The Arab garrisons of Iraq supported the pious warrior Ali, cousin and son-in-law of Mohammed. Outmaneuvered in prolonged negotiations, he was assassinated in 661 as he stood at the door of a newly completed mosque on the Euphrates, in Kufa. Nineteen years later, his son Hussein, living quietly in Medina, was persuaded by his partisans in Iraq to renew his family's claim to the caliphate. Uncertain of his exact plans, Hussein set off across the desert in A.D. 680 with seventy-two members of his family and his retainers. When they got to Kufa, they found they had been betrayed. Hussein's supporters had been rounded up, and there was no local uprising to support him. Ubaidullah, the governor sent by Yazid, the caliph in Damascus, surrounded the little band with four thousand archers and cavalrymen and demanded total surrender.

In their last stand, Hussein and his supporters dug a ditch behind them to cut off their own retreat and show their determination not to surrender. As they began to fall under the arrows of Ubaidullah's archers, Abbas, the heroic warrior-brother of Hussein, heard the women and children in their party calling out for water. He fought his way through to the Euphrates with waterskin in hand, but as he returned from the river his hand was hacked off. Abbas, whose picture as a mailed warrior going to war often hangs on the wall of Shia houses in Iraq, propped himself against a palm tree, where his enemies bludgeoned him to death with clubs and branches. Hussein himself was the last to die, his sword in one hand and the Koran in the other. The death of the two brothers in battle became the founding myth of the Shia faith. They were buried in Kerbala, where their tombs, along with that of Ali at Najaf fifty miles away, became the chief shrines of Shia Islam, attracting pilgrims from across the Islamic world. The last battle, with its theme of betrayal, suffering, martyrdom, and redemption, has the same significance in the Shia tradition as the crucifixion of Jesus in Christianity. It appeals to the downtrodden and has always created doubts about the legitimacy of rulers in Baghdad. The shrine-tombs of Hussein, Abbas, and Ali became, for the Shia, the equivalent of Mecca and Medina. Rulers of Iraq, from the Ottomans to Saddam Hussein, have had to cope with the fact that the holiest shrines of Shiism, the dominant religion of Iran and with millions of adherents across the Islamic world, are on their doorstep.

Deep religious differences were containable under the Ottoman Turks, who captured Baghdad in 1534 and held it for almost four hundred years. They were Sunni Muslim, but they had little control outside the cities. Tribal federations, who openly despised and flouted government authority, controlled the countryside. The tribe gave protection and a sense of identity, which the government did not. The attraction of tribal loyalty for an individual Iraqi was stated clearly by a deputy from Baghdad to the Ottoman parliament in 1910. He said: "To depend on the tribe is a thousand times safer than depending on the government, for while the latter defers or neglects oppression, the tribe, no matter how feeble it may be, as soon as it learns that an injustice has been committed against one of its members, readies itself to exact vengeance on his behalf." Governments in Baghdad grew stronger over the rest of the century, but belief in the clan and the tribe as the only true protector of the individual never died.

Under the Ottoman Turks, Iraq was not a single country. It was divided into three provinces, based in Mosul, Baghdad, and Basra. This was all about to change. In 1914, at the beginning of World War I, the British landed a small force at the southern tip of Iraq to defend the nearby Persian oil fields against possible Turkish attack. British troops easily captured Basra and, in April 1915, overconfident due to the lack of Turkish opposition, decided to push on to Baghdad. On the map, this looked deceptively easy. The Mesopotamian plain is flat, but is bisected by salt marshes and waterways, abandoned or in use, which provide ready-made defense works for a defending army. What followed was one of the most disastrous campaigns in British imperial history. The British force, under Major General Charles Townshend, marched up the Tigris to within twenty-five miles of Baghdad. At Ctesiphon, in sight of the famous brick arch of a sixth-century Persian banqueting hall, the British won a victory but suffered heavy losses against the reinforced Turks. They fell back downriver to

Kut, then as now a tumbledown and evil-smelling town in a bend of the Tigris. Here General Townshend withstood a Turkish siege of 146 days while British forces based in Basra fought desperately to relieve him. The British commander suffered a nervous collapse, which led him systematically to underestimate his supplies. Forced into launching premature attacks, the army outside Kut had lost twenty-three thousand dead or wounded by the time Townshend surrendered. A further seven thousand British prisoners died in a waterless march north to forced labor in Turkey, being treated with exceptional cruelty by the inhabitants of Tikrit when they passed through that town. Today the British cemetery, a little below the level of the Tigris in the center of Kut, has turned into a swamp, the tops of the gravestones just poking out of the slimy green water.

The next British advance was more calculated and successful. An immense base was built at Basra and supplies poured in from India. In 1917, Major General Sir Stanley Maude captured Baghdad before dying of cholera. The British had always intended to annex the Turkish provinces centered around Basra and Baghdad. They were more ambivalent about taking Mosul province, which looked toward Syria and Turkey. It was the heartland of the Sunni Muslims and had a large Kurdish population. At first the British concocted a Machiavellian plan to hand over Mosul to France as part of the carve-up of the Ottoman Empire, which was to create the new Middle East. The British move, which surprised the French, was wholly self-serving. They also planned to hand over eastern Turkey to Russia and wanted the French as a cordon sanitaire between themselves and the new Russian possessions. In any event, the Bolshevik revolution in 1917 negated such arrangements with the czar. After much discussion over the next three years, the British decided that they needed the province of Mosul to defend Baghdad and Basra. They would keep all three of the Turkish provinces for themselves, thus creating modern Iraq.

It was not an idea without its opponents. From the very beginning, farsighted British officials like Captain Arnold Wilson, the British civil commissioner in newly captured Baghdad, believed the creation of the new state was a recipe for disaster. It involved welding together Shia, Sunni, and Kurds, three groups of people who detested each other. In 1919, he told the British government that the new state could only be "the antithesis of democratic government." This was because the Shia majority rejected domination by the Sunni minority, but "no form of government has yet been envisaged which does not involve Sunni domination." The Kurds in the north, whom it was now intended to include in Iraq, "will never accept Arab rule." Wilson pointed out that three quarters of the population was tribal and unused to obeying any government. These suspicions of central authority ran deep. On the eve of World War I, a tribe on the Euphrates had a chant that stigmatized the government in Baghdad as "a flabby serpent which has no venom; we have come and have seen it. It is only in times past that it kept us in awe."

Two years after Great Britain drove the Turks from the provinces that were to become Iraq, the country was on the verge of the greatest revolt in its history before 1991. It broke out in July 1920 among the tribes of the middle and lower Euphrates, but had support in other parts of the country. As with the uprising seventy years later, it caught most experts on Iraq by surprise. Gertrude Bell, then the most famous British traveler and writer in the Arab world, was posted as an adviser to the British authorities in Baghdad. Just as the first shots of the rebellion were fired, she was assuring the newly arrived British military commander General Aylmer Haldane that all was well. She told him that, having conducted many "heart-to-heart interviews" with her Iraqi contacts, she believed: "The bottom seems to have dropped out of the agitation and most of the leaders are only too anxious to let bygones be bygones."

The rebellion was essentially tribal, but the British had managed to offend almost every section of Iraqi society during their brief occupation. During the war with the Turks, the British had promised an Arab regime, but had not delivered. An Iraqi notable who saw Gertrude Bell just before the uprising in 1920 told her: "Since you took Baghdad, you have been talking about an Arab government, but three years or more have elapsed and nothing has materialized." There were other causes of Iraqi resentment. Officers and officials who worked for the Turks were marginalized by the British. The Shia clergy disliked the new authorities because they were Christians. The tribes resented them because they were more diligent than the Turks in collecting taxes. The tribesmen were also, as the British were to discover, heavily armed with modern rifles that had come into their hands during the war. When the fighting was over, the British confiscated sixty-three thousand of them. In Baghdad, nationalists demanded self-determination. Gertrude Bell, who alternated between nervousness and overconfidence, railed against agitators in the city calling for unity of Islam and independence for Iraq. She wrote: "They have created a reign of terror; if anyone says boo in the bazaar, it shuts like an oyster."

The revolt lasted into 1921. It was far more serious than anything the British had expected. By the time it was over, they had lost 2,269 dead and wounded and the Iraqis an estimated 8,450. Tribes in the mid-Euphrates region ambushed and almost wiped out a battalion of the Manchester regiment. The rebels made skillful use of their rifles, though they were short of ammunition. "The Arab is most treacherous," concluded General Haldane in frustration in his notes on fighting guerrillas. "He will overpower a small detachment, and when a larger force appears he will put up white flags and be found working peacefully in his fields—incidentally, with his rifle within easy reach." The rebellion was never likely to succeed, but it provided a potent myth for Iraqi nationalists. The uprising also saw the first tentative move toward unity between Shia and Sunni, who held their first joint religious ceremonies in centuries. However much they disliked each other, some, at least, hated the British more.

The British plan was to rule Iraq at one remove and cheaply, through an Arab king. The problem with this quasi-colonial control was that a monarch appointed by the British, whose very name was unknown to the Iraqis, was tainted from the beginning. Different candidates for the throne of Iraq were considered. The final choice in 1921, backed by Gertrude Bell and T. E. Lawrence, fell on Faisal, third son of the leader of the powerful Hashemite family, Hussein of Mecca, whose claims rested on his participation in the revolt against the Turks. The British distanced themselves from the immediate problems of ruling the country, but it remained under their effective control. In a poll, the results of which eerily resemble more recent Iraqi elections, the government in Baghdad announced that 96 percent of Iraqis had voted for Faisal I, the only candidate, to be king of Iraq. A hint of the monarch's real relationship with Great Britain came a few years later when Faisal toured England. His tailor and his perfume supplier in London both asked the Colonial Office to underwrite his bills, which he had failed to pay after a previous visit.

Great Britain did not rely on Faisal I alone to rule Iraq. But they wanted to reduce the cost of maintaining a garrison, and the solution was to use air power. This has always seemed an attractive option in Iraq, in the 1920s as in the 1990s. The plains, deserts, marshes, and mountains of Iraq are difficult to police from the ground. The Royal Air Force was effective during the uprising. Iraq now became the testing ground for the RAF as the military backup for Faisal I and his successors. Ground troops were withdrawn. Great Britain had promised the Kurds self-determination, but eventually gave priority to incorporating them into Iraq. Arthur "Bomber" Harris, who led the British bomber offensive against Germany twenty years later, did not even pretend that he aimed for military targets. In 1924, he said: "They [the Arabs and Kurds] now know what real bombing means, in casualties and damage; they know that within forty-five minutes a full-sized village can be practically wiped out and a third of its inhabitants killed or injured." Delayed-action bombs were used. Other British officials were less confident that the bombing of civilians was an effective way of winning the hearts and minds of Iraqis. "If the Arab population realize that the peaceful control of Mesopotamia ultimately depends on our bombing women and children," wrote Sir Laming Worthington-Evans, the British secretary of state for war in 1921, "I am very doubtful if we shall obtain the acquiescence of the fathers and husbands of Mesopotamia." Not everybody was so discriminating. After the revolt of 1920, T. E. Lawrence wrote to the London *Observer* to say: "It is odd that we do not use poison gas on these occasions."

The new Iraqi government was designed to be weak. Between the proclamation of Faisal I as king in 1921 and the overthrow of the monarchy in 1958, it never established its nationalist credentials. Real power remained in the hands of a small coterie of former Ottoman officers who had fought with the British in the war. They were joined by some members of the Iraqi establishment who had stayed loyal to the Turks. For almost forty years, the same leaders followed each other in and out of power. Nuri al-Said served as prime minister fourteen times before he was killed, dressed as a woman, trying to flee Baghdad in 1958. Faisal had no illusions about the strength of his own government, which, as he admitted in a confidential memo in 1933, was "far and away weaker than the people." In the country at large, there were "more than one hundred thousand rifles whereas the government possesses only fifteen thousand." He concluded:

"There is still—and I say this with a heart full of sorrow—no Iraqi people, but unimaginable masses of human beings devoid of any patriotic ideas, imbued with religious traditions and absurdities, connected by no common tie, giving ear to evil, prone to anarchy, and perpetually ready to rise against any government whatsoever."

Faisal I did not mention the other reason for his government being so weak. It was devised by the British and backed up by Royal Air Force squadrons based in Basra and Habbaniya, northwest of Baghdad. If there were any doubts about the monarchy's dependence on Great Britain, they were laid to rest in 1941. Rashid Ali, a former Ottoman officer, became prime minister, backed by four army colonels. Encouraged by Hitler's victories in Europe, the new government sought to whittle away British imperial control. The regent, Abd al-Ilah, and Nuri al-Said were forced to flee. Great Britain sent troops from Jordan and India. Despite the rebels' hopes, German support never came and Iraqi troops were defeated after a month's fighting. The regent returned and the four colonels who had overthrown him were hanged.

The monarchy had been saved for the time being. But it depended on Britain at a moment when the British empire was being swept away. Arab nationalist army officers toppled governments in Egypt and Syria. Unlike these countries, Iraq had oil. It had been discovered in Kirkuk in 1927. From 1951, when the international oil companies wanted to punish neighboring Iran for nationalizing its own oil industry, Kirkuk started to bring in significant oil revenues. In the long term, the possession of immense oil fields strengthened authoritarian government in Iraq, as it did throughout the Middle East. Oil revenues made the state independent of society. It could pay for large armies and security forces without relying on taxes or foreign subsidies. But this came too late for the Hashemite dynasty. On July 14, 1958, troops led by Brigadier Abd al-Karim Qassim, a thin-voiced, intense, ascetic army officer of mixed Sunni-Shia background, stormed the royal palace in Baghdad. Artillery set fire to the top story of the building. As the young King Faisal II, together with the regent and the rest of the royal family, tried to escape out the back of the burning building, they were confronted by a semicircle of officers who shot them down with their submachine guns.

The fall of the monarchy ushered in a ten-year period of military coups, countercoups, and conspiracies. The price of failure increased by the year. Qassim was overthrown and killed in a bloodbath in 1963 in which five thousand were slaughtered, many of them after being tortured. It was the height of the cold war. The United States became increasingly involved after the overthrow of the British-backed monarchy. In 1959, Allen Dulles, the director of the CIA, told the Senate Foreign Relations Committee: "Iraq today is the most dangerous spot on earth." The monarchy was weak, but its successors were even weaker. The Arab nationalism of the new leaders seemed like a mask for Sunni Arab domination to the Iraqi Shia and had no appeal for the Kurds. Kurdish nationalism was growing in strength and its leaders were soon in a semipermanent state of rebellion.

Saddam Hussein al-Tikriti, who was to determine the fate of Iraq for most of the second half of the century, came of age at a critical moment in the history of Iraq. He was twenty-one when the monarchy was overthrown. Over the next decade, he learned the bloody mechanics of Iraqi politics. By 1968, he showed that he understood them perfectly. When he was only thirty-one years old, he helped engineer the two coups, within two weeks of each other, in which the Arab nationalist Baath Party, led largely by men from his home district of Tikrit, seized power. The political musical chairs of the previous ten years ended. Saddam and his party are still in power thirty years later. In later years, Saddam liked to portray himself as a man who succeeded in the face of adversity. By the 1980s, Iraqi poets were winning prizes for drawing parallels between Saddam and the Prophet Mohammed, both of whom were orphaned at an early age. In reality, Saddam came from a Sunni Arab family with just the right connections to propel him to the front of Iraqi politics.

He was born in Ouija, a typical Iraqi village of mud-brick houses, in the plains of northern Iraq on April 28, 1937. His father, Hussein al-Majid, was a peasant farmer who died either just before Saddam was born or a few months afterward. He was brought up by his mother, Subha al-Tulfah, a strong-looking woman who invariably wore the dark robes of the Iraqi countryside, and his two uncles. One was his mother's brother, Khairallah Tulfah, who lived mainly in Baghdad. He was not only Saddam's uncle and foster parent but also his prospective father-in-law, since he and his sister Subha arranged for Saddam, when he was five, to marry Khairallah's daughter Sajida. The marriage took place in 1963, when Saddam returned from exile in Egypt. Photographs taken when he was still a boy illustrate the real dynamics of the family better than myths subsequently woven by critics or propagandists. They show individuals from a traditional society trying to master the modern world. Khairallah Tulfah, living in the city, has neatly parted hair, but looks uncomfortable in a tie, white shirt, and checked jacket. Saddam's stepfather, Ibrahim al-Hassan, who stayed in Ouija, wears a white headdress and a traditional long robe, and carries a double-barreled shotgun by his side.

The strength of Saddam's family and clan connections matter because he was born into a tribal society. He has maintained many of its characteristics throughout his life. It was a world of intense loyalties within the clan, but cruel and hostile to outsiders. "Myself and my cousin against the world," says an old Arab proverb. Saddam later painted a picture of a deprived childhood, claiming his stepfather would rouse him at dawn by saying: "Get up, you son of a whore, go tend the sheep." His critics also stressed early traumas to prove that he came from a dysfunctional family. In fact, his reliance on his half-brothers—Barzan, Sabawi, and Watban—and his cousins, like Ali Hassan al-Majid, to stock the senior ranks of his regime argue that his inner family was always tightly knit against the outside world, whatever its inner tensions.

Saddam came from the al-Bejat clan, part of the Albu Nasir tribe, which was strong in and around the nondescript town of Tikrit, on the Tigris a hundred miles north of the capital. Set on low bluffs above the

river, Tikrit was a decayed textile town, once known for building rafts
to carry melons to Baghdad. In so far as it was famous for anything, it
was as the birthplace in the early twelfth century of Saladin, the Arab
hero, though of Kurdish background, who defeated the Crusaders.
Otherwise, Tikrit made little mark on Iraqi history. Its inhabitants
were Arab Sunni with a curious reputation for being long-winded. "To
talk like a Tikriti" is an Iraqi saying meaning to be too garrulous. By the
time Saddam was growing up, the town no longer depended on trade
and agriculture alone. Its young men increasingly took the road to
Baghdad to get jobs in the government and, above all, in the army.
Few of the sons of established families in the capital were joining its
officers' corps. Shia and Kurds had little loyalty to the state. It was
young men, often the sons of petty tradesmen and landowners, from
provincial towns like Tikrit on the upper Tigris and Euphrates, who
saw the army as a route to power.

"One of my uncles was a nationalist, an officer in the Iraqi army,"
Saddam later recalled in a rare interview about his background. "He
spent five years in prison after the revolution of Rashid Ali Kaylani [in
1941]." Saddam was only four when his uncle, Khairallah Tulfah, was
jailed, but he says that he often asked his mother what had happened
to him. She would reply: "He is in prison." Other relatives also reached
important positions in the army. One of them, Ahmed Hassan al-Bakr,
a reserved, quiet-spoken, but very ambitious brigadier, had a critical
influence on Saddam's career. He was one of the rebel officers who
took part in the overthrow of the monarchy in 1958 and later quarreled
with Qassim. He was born in 1914 into a family of petty notables who
traditionally produced leaders for the Bejat clan. He was a leader of
the 1963 coup, after which he became prime minister. Given that Iraqi
politics at this time were dominated by the military elite, Saddam, who
never entered the army, could only have risen to power in tandem with
a senior military officer.

Nor was it only in the upper ranks of the officers' corps that Saddam found allies and sympathizers. Sunni Arabs were only a fifth of
the Iraqi population, but since Ottoman times they had found jobs
as petty officials. An example of the usefulness of this for the young
Saddam was his astonishingly good treatment in the different prisons in which he was later incarcerated for political activity. For most

Iraqis in the 1950s and 1960s, these were places of torture and fear. But in 1959, by his own account, Saddam was arranging for local members of the Baath Party to be jailed with him in Tikrit because it was safer for them in prison than on the streets. In another prison in Baghdad in the 1960s, a Communist was being tortured for sawing through the bars of his cell. Saddam went to the prison governor and said he had cut the bars himself. Nothing happened to him. At a critical moment for the young Baath Party leader, in 1966 he escaped from jail on his way back from the Higher Security Court, where he was on trial for trying to overthrow the regime. His plan had been to enter the presidential palace and machine-gun the government leaders attending a meeting. Despite the seriousness of the offense, he persuaded his prison guards to take him to a restaurant in Abu Nawwas Street, where Iraqis eat fish beside the Tigris, on his way back from court. During the meal, he and six companions simply walked out the back door of the restaurant.

His family may not have been rich or powerful in the 1940s, but they knew who they were. Saddam himself, in what sounds like a truthful explanation of his own social background, said he became a nationalist and not a Communist because in central Iraq, where he came from, social divisions were not great. He contrasted this with the south and Kurdistan, where there were great landed estates. "I never felt at a social disadvantage, even I, a peasant's son," he said. He said the biggest landowner in the district was a relative of his cousin Ahmed Hassan al-Bakr. "If he got angry, he beat his relatives, but they gave him back as good as they got. As a matter of fact, they beat him much more than he beat them."

The Iraqi countryside was a violent place in which everybody carried firearms. At first the family wanted Saddam to be a farmer, but at the age of eight, Adnan, his cousin, the son of Khairallah Tulfah and later Iraq's defense minister, told Saddam that he was learning to read and write in Tikrit. Saddam was unable to persuade his family to let him go to school. One day before dawn he set off across the fields to make his own way there. On the road he met some relatives, who approved of his educational plans and agreed to help him. Their response underlines the degree of insecurity in provincial Iraq in the 1950s. "They gave him a pistol and sent him off in a car to

Tikrit," says his official biographer. Accounts of Saddam's early bloodthirstiness are suspect, but Dr. Abdul Wahad al-Hakim, an Iraqi exile, says Saddam was quite prepared to use his fearsome reputation in the next few years. He recalls:

"My headmaster told me that he wanted to expel Saddam from school. When Saddam heard about this decision, he came to his headmaster's room and threatened him with death. He said: 'I will kill you if you do not withdraw your threat against me to expel me from the school.'" At the age of ten, Saddam went to stay with Khairallah Tulfah in Baghdad, but with frequent trips home to Ouija and Tikrit.

Later, after Saddam's ascent to power, "Tikritis" was to become a nickname for the Iraqi political elite. But after the overthrow of the king in 1958, Tikrit was intensely and violently divided between Communists and nationalists such as Saddam. This is the background to the first killing by Saddam for which there is reliable evidence. The victim was Haji Sadoun al-Tikriti, a warrant officer and Communist leader in the city. It happened in 1959, and the dead man was said to be a distant relative of Saddam's. Twenty years later, Saddam, by now vice chairman of the Revolution Command Council, came to the school of a relative of the dead man in Baghdad. Following tribal tradition, he gave him blood money and a Browning pistol.

Saddam joined the Baath Party when he was twenty, the year before the overthrow of the monarchy. Founded in Iraq in 1952, it was small and tightly organized in cells of three to seven members. Its ideology combined intense Arab nationalism with woolly socialism. Its will to power always exceeded its idea of what to do with it. Hanna Batatu, the great Iraqi historian of these years, writes: "A Baathi would have looked in vain through the whole literature of his party for a single objective analysis of any of the serious problems besetting Iraq."

But there was nothing vague about how the Baath Party intended to deal with its enemies. It had quarreled with Qassim immediately after he took power because of his opposition to pan-Arab unity with Egypt and Syria. In their first independent initiative, the Baathists decided to assassinate him. Among those recruited for the attempt was the hitherto unknown party militant Saddam Hussein, by now a law student in Baghdad. What happened next became part of Saddam's personal mythology, the topic of a government-sponsored novel and a film, *The Long Days*. In the cinematic version of the assassination attempt, the part of Saddam is played with verve by Saddam Kamel, his cousin and namesake, who somewhat resembled the Iraqi leader.

The assassination attempt on October 7, 1959, came close to success. Qassim was driving to a reception at the East German embassy. The Baath Party had a source inside the Defense Ministry who could tell them when Qassim would drive down al-Rashid Street, then Baghdad's main thoroughfare, with its white colonnades and luxury shops. Saddam's role was to provide covering fire for the four men who were to kill Qassim. Two of the gunmen were to open fire on anybody in the backseat while two aimed at the front. When the shooting started, Saddam became overexcited and drew the submachine gun he was hiding under a cloak given him by Khairallah Tulfah, his uncle. The assassins killed Qassim's driver, seriously wounded an aide, and hit Qassim himself in the shoulder. He was rushed to the hospital in a passing taxi. One of the attackers was shot dead, apparently by a chance shot from his own side. Saddam himself was hit in the fleshy part of his leg. "It was a very superficial wound to the shin," said the doctor who treated him. "A bullet just penetrated the skin and it stopped there in the shin of his leg. . . . During the night he cut it by using a razor blade and took the bullet out."

Years later Saddam told King Hussein that he had thought he would die after his failed attempt to kill Qassim. He gave lengthy and detailed accounts of his escape from the police, up the Tigris from Baghdad to Ouija. It was a critical element in his self-image as an Arab hero. For seven years after the Gulf War, Saddam was seldom seen in public. When he did reemerge, almost his first visit was to the village of al-Dhour, on the Tigris, where thirty years before he had swum ashore, hungry, his teeth chattering from the cold, and on the run.

Even if embellished, the story of Saddam's escape is a dramatic tale. The journey was long because Saddam was not able to hire a car in Baghdad. Instead, he bought a horse. Dressed as a bedouin, he rode north for four nights. When he fell into a trap laid for smugglers by police officers, he explained his lack of papers by telling them: "Bedouin do not carry identity cards." He needed to cross the

Tigris and offered to pay the owner of a barge one and a half dinars to ferry him and his horse across. The barge owner refused because of a curfew. Saddam decided he would leave his horse and swim. The water was cold and he was in a state of collapse by the time he reached al-Dhour, on the far side of the Tigris. "It was like you see in the movies, only worse," he recalled later. "My clothes were wet, my leg was injured, and I hadn't eaten properly for four days." He staggered into a house where people first thought he was a thief, but finally let him stay the night and lit a fire to dry his clothes. When Saddam tried to leave in the morning, the man of the house confronted him, saying: "Where do you think you are going? You have just swum across the Tigris with your clothes on. This means something is very wrong and we are not going to let you go until we know what the truth of the matter is." Saddam's response was to hint at tribal vengeance. He said: "Supposing I have committed a crime against a clan on the other side of the river; suppose they follow me here and kill me in your home. What good will it do you when my clan finds out that I was killed among you?" The man replied, "What you say is true. God protect us." Saddam walked on until he found one of his brothers, who was a guard at an elementary school. He reached Ouija and safety and later escaped to Syria.

The three years Saddam spent in Damascus and Cairo were the only time in his life he lived abroad. Once he gained power, his visits to foreign countries were fleeting. Most of his time in exile was spent in Cairo under the protection of President Nasser, who was at odds with Qassim. Accounts of his behavior differ. Abdel Majid Farid, the secretary general of the Egyptian presidency, who had been an Egyptian military attaché in Baghdad until expelled, says: "We helped him go to the faculty of law and tried to get him an apartment. He was one of the leaders of the Iraqi Baath. He used to come to see me now and then to talk about developments in Baghdad. He was quiet, disciplined, and didn't ask for extra funds like the other exiles. He didn't have much interest in alcohol and girls."

This sounds a little too good to be true. Hussein Abdel Meguid, the owner of the Andiana Café, where Saddam used to meet with friends in the early 1960s, describes him as a troublemaker who did not pay his bills. "He would fight for any reason," he says. "We wanted to bar him from coming here. But the police came back and said he was protected by Nasser." Meguid says Saddam finally left owing the equivalent of several hundred dollars.

Both the presidential adviser and the café owner were to meet Saddam again. He had a highly developed sense of favors received or denied. Abdel Majid Farid was jailed by President Sadat after the death of Nasser and left Egypt to live in Algeria. Fifteen years later, he met Saddam Hussein again. He was invited to Baghdad and received financial help. At the Andiana Café, Meguid also saw Saddam after he became ruler of Iraq. He recalls, "When he was vice president in the 1970s, he came back to Cairo, and he came here. He paid his bills and three hundred pounds extra."

In early 1963, Saddam had more important things to worry about than his outstanding bill at the Andiana Café. On February 8, a military coup in Baghdad, in which the Baath Party played a leading role, overthrew Qassim. Support for the conspirators was limited. In the first hours of fighting, they had only nine tanks under their control. The Baath Party had just 850 active members. But Qassim ignored warnings about the impending coup. What tipped the balance against him was the involvement of the United States. He had taken Iraq out of the anti-Soviet Baghdad Pact. In 1961, he threatened to occupy Kuwait and nationalized part of the Iraq Petroleum Company (IPC), the foreign oil consortium that exploited Iraq's oil.

In retrospect, it was the CIA's favorite coup. "We really had the *t*s crossed on what was happening," James Critchfield, then head of the CIA in the Middle East, told us. "We regarded it as a great victory." Iraqi participants later confirmed American involvement. "We came to power on a CIA train," admitted Ali Saleh Sa'adi, the Baath Party secretary general who was about to institute an unprecedented reign of terror. CIA assistance reportedly included coordination of the coup plotters from the agency's station inside the U.S. embassy in Baghdad as well as a clandestine radio station in Kuwait and solicitation of advice from around the Middle East on who on the left should be eliminated once the coup was successful. To the end, Qassim retained his popularity in the streets of Baghdad. After his execution, his supporters refused to believe he was dead until the coup leaders showed pictures of his bullet-riddled body on TV and in the newspapers. By one account, Qassim was buried in a shallow, unmarked grave. The corpse was unearthed by dogs who began to eat it. Horrified by this, peasant farmers reburied the body in a coffin, only for the secret police to dig it up again and throw it in the Tigris.

The triumph of the Baath Party was brief. It was deeply divided between its civilian and military wings. The new prime minister was Brigadier Ahmed Hassan al-Bakr, Saddam's cousin. Many of the other senior officers who overthrew Qassim were from Tikrit, though they belonged to a tribe different from that of al-Bakr and Saddam. There was little to hold the party together other than hatred of its enemies. In November, the new president, Abd al-Salaam Aref, first persuaded the military Baathists to turn on the civilian wing of their party and its militia. Soon afterward, Aref expelled the Baathist officers from the government.

Saddam played no role in the 1963 coup. It is not even clear that he took part in the massacres afterward. The following year he was jailed, but conditions were not onerous. The debacle of the Baathists' first bid for power put Saddam and al-Bakr in charge of the party. They planned to seize power again, avoiding the mistakes of 1963. The party was not strong enough to act on its own, but they suborned Abd al-Razzaq al-Nayif, the head of military intelligence. The coup took place on July 17, 1968, and, in contrast to what had happened five years before, this time it was the non-Baathist officers who were ousted within thirteen days of taking power.

Nine years after he tried to kill Qassim, Saddam was vice chairman of the Revolution Command Council (the RCC) and the second most powerful man in Iraq. The extent of his influence was kept deliberately shadowy. He was a civilian in what was, until the late seventies, primarily a military regime, with Ahmed Hassan al-Bakr as the president. In the 1970s, Saddam tried to ensure that army officers did not see foreign publications referring to him as "the new strongman of the regime." He seems to have assumed that they would not read Baath Party documents referring to them as "the military aristocracy." After the 1968 coup, the triumphant Baathists were as bloodthirsty as five years before, but their violence was more systematic. No opponent would get a second chance. Saddam took Nayif, the military intelligence chief who had assisted the

Baathists in their coup and then been displaced, to the airport with his gun in Nayif's back. Even in exile, Nayif was considered a possible threat. In 1974, an assassin tried to kill him in his London apartment. Four years later, he was shot dead in a hotel in the same city. General Hardan al-Tikriti, the minister of defense, was dismissed in 1970 and was assassinated in Kuwait the following year. Previously, no regime in Iraq had been stable because army, party, tribe, and security services competed for power. Between 1968 and 1979, Saddam was able to get a grip on all four centers of power, which made him almost impossible to overthrow.

Saddam at this period had charm as well as ferocity. His attitude was very tribal, merciless and unforgiving to enemies, grateful and generous to friends. "There is no real mystery about the way we run Iraq," one of Saddam's associates once said. "We run it exactly as we used to run Tikrit." Saddam's rise was extraordinarily rapid. Less than ten years after he had fled for his life from Baghdad, he was the second most powerful man in Iraq. He was hardworking, sleeping little and rising before dawn. There was little wrong with his health, though in later years he had trouble with his back. He developed a taste for Portuguese Mateus rosé wine. When younger, he smoked a pipe, but President Houari Boumedienne of Algieria introduced him to cigars, which he has smoked ever since. For relaxation, all the Tikritis were fans of gypsy dancing, known as *kawliya*. Bakr used to phone Iraqi television to ask for a program of gypsy dancing. When it was finished, he would call again, congratulate them, and ask for more dancing. Iraqi viewers were irritated to find promised coverage of football matches abandoned for the *kawliya*. (When General Hussein Kamel fled to Jordan in 1995, one Iraqi said presciently: "He will go back to Iraq in the end. He can't last without *kawliya*.") Television schedules only returned to normal when Saddam became president, not because he was any less enamored of *kawliya*, but because he had bought a video machine.

Saddam replaced al-Bakr as president in July 1979. The bloodbath with which he began his rule left no Iraqi in any doubt that all authority would in future stem from him. This is important because it explains why nobody within the leadership tried to dissuade him from invading Iran in 1980 or Kuwait in 1990.

No criticism was allowed. This mattered less in terms of domestic Iraqi politics, where Saddam showed great skill. But in foreign affairs, his lack of experience and unwillingness to take advice was a recipe for disaster.

The opening moves in the crisis that led to the purge of the party were in early July 1979. President Bakr announced that he was to resign and hand over his office to Saddam at a meeting of the RCC on July 10. He said he was in poor health. But it rapidly emerged that there was strong opposition to Saddam among other leaders. Muhie Abdul-Hussein Mashhadi, the secretary of the ruling council, objected and demanded a vote on the decision. "It is inconceivable you should retire," he told Bakr. "If you are ill, why don't you take a rest?"

Saddam's opponents had waited too long to act. Muhie Abdul-Hussein Mashhadi was arrested for questioning and presumably tortured. Barzan, Saddam's half-brother, headed the investigation. In the next few days, Saddam toyed with those he was about to destroy. On July 18, party leaders were invited to a dinner party at the presidential palace. After the meal, they were each asked to write a detailed report of any meetings they might have had with Abdul-Hussein or another suspect, Mohammed Ayesh, the industry minister, the previous year. The circle of suspects increased. Barzan accused Ayesh of acting for Syria, Iraq's hated rival. In all, five members of the RCC, a quarter of its membership, were expelled. Along with sixteen others, they were executed on August 8. Branches of the Baath Party throughout Iraq each sent a delegate with a rifle to join the firing squad.

Saddam wanted the purge to create maximum terror and so ordered a videotape to be made of one of a series of meetings where he singled out those accused of conspiring against him. The tape indeed records a numbing and carefully orchestrated spectacle of terror. As it begins, the delegates to a meeting of the Baath Party leadership wait anxiously as Saddam prepares to speak.

"We used to be able to sense a conspiracy with our hearts before we even gathered the evidence," he tells the party leaders. "Nevertheless, we were patient and some of our comrades blamed us for knowing this, but not doing anything about it."

In a prearranged move, a Baath Party official gets up and admits his guilt. Others call for a purge. Ali Hassan al-Majid, the cousin of Saddam later notorious for using chemical weapons against the Kurds, says to him unctuously: "What you have done in the past was good. What you will do in the future is good. But there's this one small point. You have been too gentle, too merciful."

"Yes, that's true. People have criticized me for that," replies Saddam. "But this time I'll show no mercy." After half an hour a "conspirator" is taken away from the meeting. Nobody has any doubt about his fate.

The camera focuses on Saddam looking relaxed and smoking a cigar. Then he rises to speak again and now his voice has turned harsh: "The witness has just given us information about the group leaders in that organization," he shouts. "Similar confessions were made by the ringleaders. Get out! Get out!"

In a frenzy of fear, other members of the party leadership shout: "Long live the party! Long live the party! God save Saddam from conspirators!" As if affected by these loyal cries, Saddam begins to weep. He goes to sit among the party members in a show of solidarity. He invites them to join the firing squad that is to execute their colleagues.

Ambassadors and officials abroad suspected of involvement in the conspiracy were called to Baghdad. For the first time, the families of "traitors" were also punished. The body of one senior leader was returned to his house in Baghdad in a pickup truck. The body showed signs of torture. A note attached to the corpse said the leader had died of a heart attack and ordered that there should be no mourning.

It took a year for the significance of Saddam's takeover in Baghdad to become apparent to the rest of the Middle East. For the sixty years after Britain created Iraq, it was paralyzed by its own divisions. Despite growing oil wealth, it remained a third-rate power, unable to mobilize its resources. The purge of the Baath Party leadership in 1979 gave Saddam total control. He eliminated competitors for power within the party. He had already disposed of those outside. The long-running Kurdish rebellion, which had destabilized previous Iraqi governments, ended in 1975 when the shah of Iran withdrew his support for the Iraqi Kurds in return for territorial concessions by Iraq. The country was growing wealthier. It produced 3.4 million barrels of oil a day and, after Saudi Arabia, had the largest oil reserves in the Middle East.

Internal feuding in the 1960s and 1970s made Iraq a marginal power in the Middle East and a very small player in world affairs. Saddam now started on a sustained effort to win control of the Persian Gulf and leadership in the Arab world. His campaign had two phases: The first began with his invasion of Iran in 1980 and ended with Iraq's qualified victory eight years later. The second was much shorter. Frustrated by what he saw as an attempt by Kuwait—backed by the United States and Britain—to rob him of the fruits of his victory over Iran by driving down the price of oil through over-production, Saddam invaded the emirate on August 2, 1990. It was a venture far beyond Iraq's political and military strength. The Americans and British were never likely to allow Iraq to win control of the Gulf, which has 55 percent of the world's proven oil reserves.

In 1979, this final disaster lay far in the future. On the contrary, the political situation in the Gulf seemed to offer Iraq great opportunities. In 1979, Ayatollah Khomeini, after sixteen years in exile in the Iraqi holy city of Najaf and subsequently in France, overthrew the shah and returned to Iran. Militant students further radicalized the revolution by taking over the American embassy and holding American diplomats hostage. This had repercussions in Iraq. Militant Iraqi Shia groups saw no reason why the revolution in Tehran should not be repeated in Baghdad.

It was not an idea ever likely to succeed. Iraq has a more secular tradition than Iran. The threat of an Islamic revolution was always likely to unite the Sunni Arab core of the regime behind Saddam. Nor had Islamic fundamentalism any appeal for the Kurds, still less the small but influential Christian minority. It may not have been a coincidence that one of the first attacks by al-Dawa, the militant Islamic group, was against Tariq Aziz, then deputy prime minister, who was a Chaldean Christian from Mosul. A militant from Dawa threw a grenade at him as he was opening a student conference on April 1, 1980, at the Mustansariyah, the ancient university in the heart of Baghdad. The next day, standing in the pouring rain, Saddam told a meeting of students: "The Iraqi people is now a large and powerful

mountain they cannot shake with all their bombs. By God, the innocent blood that was shed at Mustansariyah will not go unavenged." A
few days later, a second bomb was thrown into the funeral procession
of those killed in the first attack. Vengeance followed as promised.
Mohammed Bakr al-Sadr, a senior religious leader and one of the
heads of al-Dawa, was executed, along with his sister. Thirty thousand
Iraqis of Iranian origin were expelled from Iraq. Saddam began to
refer to Ayatollah Khomeini as "that mummy" and Khomeini called for
the Iraqi army to leave its barracks and overthrow Saddam.

This was largely a blind. There was little threat to Saddam's leadership from the disarmed and leaderless Iraqi Shia. Instead, the chaos in
the Iranian army and the diplomatic isolation of Iran seemed to offer
Saddam an opportunity. A hint of the real thinking in the Iraqi leadership at this time comes from a peculiar source. A declassified note from
an agent, whose name is blacked out, of the Defense Intelligence
Agency, the intelligence arm of the Pentagon, reported from Baghdad
on April 8 that Iraq had ambitious plans for Iran that had nothing to do
with bomb attacks by al-Dawa. He said: "There is a 50 percent chance
that Iraq will attack Iran. Iraq has moved large numbers of military personnel and equipment to the Iraq-Iran border in anticipation of such an
invasion." A rocket attack on an Iranian oil field two days before had
been carried out by an Iraqi commando unit. The agent said Iraq
believed: "The Iranian military is now weak and can be easily defeated."

It was a disastrous miscalculation. The Iranian population is three
times as large as Iraq's. The Iranian revolution was popular. At first,
Iraqi tanks advanced easily, but within a year Iranian light infantry was
causing serious casualties. Iraq was disastrously defeated in the battle
of Khorramshahr. By the end of 1982, American intelligence estimated that Iraq had lost forty-five thousand dead and the same number of prisoners. There were mass surrenders of Iraqi soldiers. The
West and the Arab Gulf states worried that Iraq would collapse. They
rushed in supplies. Washington even removed Iraq from the list of
states supporting terrorism, though Abu Nidal, a terrorist leader, was
living in Baghdad at the time. Saudi Arabia gave Iraq $25.7 billion and
Kuwait $10 billion, mostly in the first two years of the war. Iranian
troops failed in their assault on Basra. They also found, once they had
crossed into Iraq, that ordinary Iraqi soldiers, mostly Shia, stopped

surrendering. Within two years, the CIA was giving Iraqi military intelligence regular briefings and satellite photographs showing Iranian positions. By 1984, the U.S. embassy had reopened in Baghdad.

Supported by the United States and the Soviet Union, Western and Eastern Europe, as well as most of the Arab world, Saddam believed he could sustain a long war. This long-haul strategy changed only when Iran captured the Fao peninsula, a desolate but strategically important triangle of shifting sand in the far south of Iraq, which sticks out into the Gulf, by a surprise attack in 1986. Iraq planned a counterattack. The Republican Guard was expanded from one to thirty-seven brigades. More weapons were needed. A problem was the low price of oil. This hit Iraqi revenues and the pockets of previously generous allies in the Gulf. Iraq looked increasingly to the United States and Britain, along with Australia the only Iraqi creditors still being paid. After a visit to Iraq, Clement Miller, an Eximbank credit specialist, reported that Iraqi officials told him not to worry "because Saddam Hussein himself has sent around a circular that said, very simply, 'Pay the Americans.'"

In the Gulf, Iraqi planes were attacking Iranian oil tankers with French Exocet missiles. The Iranians retaliated against Kuwait. By agreeing that the Kuwaiti tanker fleet could sail under the American flag, the United States effectively joined, on the Iraqi side, the so-called "tanker war" in the Gulf. The U.S. Navy attacked Iranian oil facilities, eliminated the small Iranian navy, and in July 1988, shot down an Iranian civilian airliner en route to Dubai, killing all 290 civilians onboard. Iranian President Akbar Hashemi Rafsanjani believed this showed that the United States had joined the war on the Iraqi side. He persuaded Ayatollah Khomeini that the odds against Iran were now too great. The Iranian leader, telling his people that he must "drain the bitter cup," accepted a cease-fire on August 8, 1988.

There was a second, unspoken reason for Iran ending the war. Iraq had been using poison gas extensively on the battlefield from 1984 on. On April 17, 1988, the Republican Guard counterattacked at Fao in a carefully prepared assault. The troops had previously trained at a specially built model of the battleground at Lake Habbaniya, northwest of Baghdad. Iranian Revolutionary Guards were pounded by heavy artillery, aerial bombing, and gas. After two days of fighting, the Iranian troops were routed. Iraq used not only mustard gas, but nerve gases such as tabun and sarin. The mixture of deadly gases made it impossible for the Iranians to take countermeasures against all of them. The effectiveness of gas attack—and the failure of the outside world to react—may have convinced Saddam of the great importance of the weapon. This explains his determination not to give it up.

The war with Iran made Iraq a regional power and the strongest of the seven countries that border the Gulf. It started the war in 1980 with an army of ten divisions and ended it with fifty-five divisions. By the end of the war it had a tank force of four thousand and rockets that could reach Tehran or Tel Aviv. By surviving initial military defeats, Saddam proved the durability of his regime. For all the government's patriotic exaggerations, the Iraqi Shia had fought fiercely against their coreligionists in Iran. Iraq won support from both superpowers, the Europeans, and much of the Arab world.

None of this was cost-free. Iraq, with a population of just 17 million people, ended the war with at least 200,000 dead, 400,000 wounded, and 70,000 taken as prisoners. It is difficult today to find an Iraqi who did not lose a close relative. There was also the financial cost. Saddam fought the war on credit. By the end of the war, he owed $25.7 billion to Saudi Arabia, $10 billion to Kuwait, and smaller sums to other Arab states. A further $40 billion was owed to the United States, Europe, and the rest of the industrialized world. Saddam later made much of Iraq's economic crisis because of the war. Confronting the emir of Kuwait at an Arab summit conference in Baghdad in April 1990, he said: "War doesn't mean just tanks, artillery, or ships. It can take sub-tler or more insidious forms, such as the overproduction of oil, economic damage, and pressures to enslave a nation." This sounds exaggerated. Iraq's oil revenues in 1990 were due to rise to $13.7 billion in the course of the year. It was in a much better condition to pay its debts than countries like Brazil or Argentina.

The victory over Iran was real, but Saddam grossly exaggerated its extent. This was underlined on August 8, 1989, a year after the end of the fighting, when an extraordinary monument was opened in Baghdad. It was an Iraqi Arc de Triomphe. Two metal forearms, each forty feet long, reach out of the ground clutching steel sabers, whose tips cross, forming an arch under which the Iraqi army passed. The arms were modeled from a cast taken of Saddam's own arms. They were too big to be made in Iraq and were cast in a metal foundry in Basingstoke, England. The invitation to guests for the inauguration of the monument captures the flavor of the event:

"The ground bursts open and from it springs the arm that represents power and determination, carrying the sword of Qadissiya. It is the arm of the Leader President Saddam Hussein (God preserve and watch over him) enlarged forty times. It springs out to announce the good news of victory to all Iraqis and pulls in its wake a net that has been filled with the helmets of the enemy."

Did the U.S. and Britain move to limit Iraq's power in the Gulf after the end of the war? Iraq later published a report from Brigadier Fahd Ahmed al-Fahd, Kuwaiti director-general of state security, about his visit to the CIA in October 1989. One item says: "We agreed with the American side that it was important to take advantage of the deteriorating economic situation in Iraq in order to put pressure on that country's government to delineate our common border." The point is only one of many in the report, and it is not entirely surprising that Kuwait thought this a good moment to settle a territorial dispute concerning Bubiyan and Warba, two Kuwaiti islands that block Iraq's access to the Gulf.

From February on, Iraq's relations with the U.S. and Britain deteriorated rapidly. In 1989, Saddam severed relations with the CIA. John Kelly, the U.S. assistant secretary of state, visited Baghdad. He said: "You are a force for moderation in the region, and the United States wants to broaden her relationship with Iraq." But Saddam immediately took exception to criticism of himself on the Voice of America. On February 23, Saddam warned Arab leaders meeting in Jordan of the waning power of the Soviet Union and the growing dominance of the United States in the Middle East. He said: "If the population of the Gulf—and the entire Arab world—is not vigilant, this area will be ruled by the United States." At the same time, he arrested Farzad Barzoft, an Iranian-born journalist working for the British newspaper the *Observer*, who was accused of espionage. King Hussein went to Baghdad to appeal for his release. Abdul Karim al-Kabariti, who later became the Jordanian prime minister, recalls: "The King said to Saddam, 'This is the beginning of the slippery slope [toward Iraq breaking with the West]. Do not kill him even if he is a spy.' Saddam said: 'I'll see what I can do about it.' When the king got back to Amman, he found that Barzoft had been executed [on March 15]." Saddam systematically escalated the crisis. On April 2, Baghdad radio broadcast a speech he made to army officers. He said: "If the Israelis try anything against us, we'll see to it that half their country is destroyed by fire." The speech was in strong Iraqi dialect and designed for a domestic audience. "I shall burn half your house" is a colloquial expression common among Baghdad street toughs.

It was a measure of Iraq's strength that on April 28 it could still get twenty-one Arab monarchs and heads of state to attend a summit in Baghdad. Here Saddam targeted Kuwait for waging economic war against Iraq. By the middle of July, the first Republican Guard division had moved to the Kuwaiti border. Two more followed. There was an atmosphere of crisis, but an underlying presumption that the worst Iraq could do would be to settle its border dispute with Kuwait by force. This explains why April Glaspie, the U.S. ambassador in Baghdad, in a notorious interview with Saddam on July 25, emphasized that the United States had no opinion on "your border disagreement with Kuwait." Five days later, John Kelly told the Middle East subcommittee of the House of Representatives in Washington that there was no obligation for the United States to use its forces if Iraq invaded Kuwait.

Kuwait certainly believed the threat was limited. Saddam drove its emir, Sheikh Jabr al-Sabah, to the airport after the Baghdad summit and asked for the use of the disputed islands. The emir said he could not give up part of his country. On July 31, on the eve of the Iraqi invasion, Sheikh Jabr wrote to his brother telling him to make no concessions to the Iraqis at a final summit in Jeddah. "The Saudis want to weaken us and exploit our concessions to the Iraqis so that we will make concessions to them in the demilitarized zone," he wrote. "As for the Iraqis, they wish to compensate for the cost of their war from our resources. Neither demand will bear fruit . . . that is also the position of our friends in Egypt, Washington, and London."

Most of the Iraqi leadership probably realized that Saddam had overplayed his hand, but after 1979 they were unlikely to contradict him. Tariq Aziz, the suave Iraqi foreign minister, later disclosed that the original Iraqi plan was for a partial invasion of Kuwait. The Iraqi army was to seize Bubiyan and Warba, as well as the disputed Rumailah oil field that straddles the border between the two countries. The decision to take the whole of Kuwait was taken "at the last moment" by Saddam himself. He argued that "it would make no difference to the United States" how much of Kuwait was taken by Iraq.

It made all the difference in the world. Saddam could have gotten away with taking two barren islands off Kuwait, but not the whole emirate. The U.S. and Britain were never going to hand over hegemony in the Gulf to Iraq. By taking the whole of Kuwait, Saddam made easy their task of uniting the rest of the world against him. It was, perhaps, one of the greatest political miscalculations by any leader since Hitler invaded the Soviet Union in 1941. Perhaps at the last minute, the irrational element in his personality, the warrior-hero, came to the fore as he compared himself with Nebuchadnezzar, Sargon of Akkad, the Prophet Mohammed, and Saladin. At the height of the Iran-Iraq war he had rebuilt Nebuchadnezzar's palace in the ancient city of Babylon on the Euphrates. Now he would establish himself as the preeminent Arab leader who broke the power of the West and what he termed "the emirs of oil" in the Middle East. In any event, once the die was cast, there was to be no compromise or retreat.

In the six months between the invasion of Kuwait and the start of the allied bombing, Iraqis waited for Saddam to pull out of Kuwait. Once the United States had assembled its vast coalition and was building up its army in Saudi Arabia, Iraq's position got weaker and more isolated. Saddam had sympathy on the street in many Arab countries, but there were no revolutions. The Soviet Union cooperated with the United States. At the end of the day, Saddam's own army would not fight. Eleven years after Saddam began his campaign to make Iraq a regional superpower, the country had been reduced to semicolonial servitude.

Nothing illustrated the Iraqi leader's humiliation more starkly than the enemy he was forced to tolerate in his midst. Years before, he had determined to build himself weapons that would force his neighbors—and the rest of the world—to acknowledge his power. Now the victorious allies had made him accept a group of officially sanctioned spies, charged with rooting out the weapons and the secrets that surrounded them.

FOUR

Saddam Fights for His Long Arm

·

Dr. Hussain al-Shahristani lay on the soft carpet, unable to move. Eight months before, in September 1979, he and the other members of the Iraqi Atomic Energy Commission had been summoned by Saddam Hussein to a special meeting. The dictator, only recently installed as the absolute master of Iraq, informed them that the country's nuclear research should be redirected to "develop our potential in strategic fields." Al-Shahristani, an internationally respected expert on neutron activation, had been the only person to object to what was obviously a plan to develop nuclear weapons. "We have signed the nonproliferation treaty and we cannot engage in nonpeaceful uses of atomic energy," he had said flatly.

"Dr. Shahristani, I suggest you stick to your field and leave politics to me," replied the dictator, who then delivered his little homily on politics being the art of saying one thing, intending to do another, and then embarking on yet another course of action.

During this exchange, the room had gone deathly quiet. Everyone at the meeting knew what was likely to happen to the short, bespectacled physicist for daring to defy Saddam. They were right.

In one of the interrogation centers of the Amn al-Amm, an intelligence service reporting directly to the presidential palace, Dr. al-Shahristani's wrists were tied behind his back by a rope that held him hanging in the air while they beat him for twenty-two days and nights. Then he was taken to face a "revolutionary court," charged with the capital crimes of spying for the United States, Iran, and Israel. There were three judges, two of whom were fast asleep.

True to form, al-Shahristani, even though half paralyzed from the torture, had lashed out at this hanging court. He was descended from the Prophet, he pointed out, who in turn was descended from Abraham. "I challenge you and the president to tell me who his grandfather was." At the sound of this suicidal insolence, the two slumbering judges woke up. "If your family has lived for five thousand years in this land," continued the defiant prisoner, "it does not matter if you respect the president or not."

The charges themselves, let alone the allusions to Saddam's parentage, were enough to earn a speedy death sentence. But apparently someone on high believed that the scientist's nuclear expertise might still be useful. He was consigned to life in prison and taken to the "special section" of the Abu Ghraib prison, where anywhere from forty to sixty men were crammed into small, windowless cells, removed only for random beatings or execution.

When they came and blindfolded him one morning in August 1980, eight months after his arrest, al-Shahristani had assumed that he was going to be killed after all. But instead, he found himself in a luxurious villa, the former residence, so he later discovered, of the minister for planning, who had been executed in the purge of the previous year. His guards showered and shaved him—his arms were still paralyzed from the torture—and drenched him in cologne before leaving him on the carpet. He was going to have some important visitors.

Two men came into the room. One of them stood by the door. Al-Shahristani recognized him as Abdul Razaq al-Hashemi, a hard-faced member of the Baathist inner circle who was then minister of higher education. The other, whose features resembled those of Saddam Hussein, came and sat on a chair close by the recumbent physicist.

This was Barzan al-Tikriti, half-brother and close confidant of the leader. At the time, he was chief of the Mukhabarat, the secret police and one of the several competing intelligence organizations deployed by the regime. When he spoke, his voice was full of friendly concern. "The president is very sorry that you have been arrested," he said. "It's all the fault of the Amn al-Amm." After cursing to Razaq at the brutal license of the rival intelligence service, he turned back to al-Shahristani. "We would like you to go back to work. We have a nice place all ready for you at the palace."

"I am physically and mentally paralyzed," replied al-Shahristani. "I am not fit to work."

"We have important programs, we need you to work on the atomic bomb. We need the atomic bomb," Barzan explained, "to give us a long arm to reshape the map of the Middle East."

"Sir . . . ," interrupted Razaq from the doorway, worried that his master was saying too much.

Barzan waved a hand to quiet him. "I mean what I say," he said, looking at the man on the floor, then turning to reassure his aide. "Don't worry, he is in our hands. He can never be free again."

Al-Shahristani tried to close off the discussion by protesting that his particular expertise would be of no use on a weapons project, but Barzan would have none of it. "We know your ability and what you can do," he said. "It is every citizen's duty to serve his country. Anyone who refuses does not deserve to be alive."

Lying on the floor, his useless arms trailing at his sides, al-Shahristani still managed to answer back. "I agree with you that we should all serve our country," he riposted. "But what you are doing does not serve our country."

Barzan looked at him, as al-Shahristani recalls, "like I was mad"—a not unreasonable assumption under the circumstances. Then, says al-Shahristani, "He gave a 'yellow smile,' as we say in Arabic, a false smile, and replied, 'At least we are agreed that we should all serve our country. Rest now and think about what I have said.'"

Al-Shahristani was to have ten years in solitary confinement to reflect on Barzan's words, but he never buckled. Eventually, during the chaos of the Gulf War in 1990, his astonishing fortitude was rewarded when he won over the "trusty" assigned to deliver his

meals. This man, a Palestinian incarcerated by Saddam as a favor to
Yasser Arafat, agreed to help him escape. Exiting the jail in a stolen
Mukhabarat car, the physicist made his way to the north and across
the Iranian border to freedom.

In al-Shahristani, Saddam and his henchmen had found a rare
individual who refused to buckle. But they had little trouble in
recruiting others to help deliver the "long arm" of which Barzan had
spoken. The nuclear program went into high gear in 1982 under the
presiding genius of Jafr Dia Jafr, al-Shahristani's friend and fellow
scientist. Jafr had tried to help his imprisoned colleague, even to the
point of telling Saddam that it would be impossible to proceed with
the project without al-Shahristani's help. The dictator interpreted
this as a threat of noncooperation and had Jafr arrested the moment
he left the presidential office. Rather than torturing this scientist,
Saddam opted to whip up his enthusiasm by having others tortured
to death in front of him. Jafr saw the light, accepted the unlimited
resources and benefits placed at his disposal, and set to work. By
1990, he was on the verge of success.

No one knows precisely how many billions of dollars were lavished
on the Iraqi bomb project. Even during the darkest years of the Iran-
Iraq war, work proceeded at full speed. The scale of the project, the
creation of a network of foreign contractors, and the success with
which the program was kept out of the international public eye were a
monument not only to the talents of Jafr and the overall director of the
scheme from 1987, Saddam's cousin and son-in-law Hussein Kamel
(Barzan had slipped from favor in 1983), but also to the insouciance of
the Western powers. It was not as if the veil of secrecy surrounding the
project was complete. Even when a close U.S. ally, Saudi Arabia,
agreed to help finance the Iraqi bomb program on the promise of
repayment in nuclear devices, Washington took no action. "I knew
about it," says one former U.S. diplomat in the region in reference to
the Saudi contribution, "and so did the CIA." In 1989, a senior official
at the U.S. Department of Energy learned that nuclear detonators of
the most advanced kind were being shipped from the United States to
Baghdad, indicating that designs for the actual operational Iraqi
nuclear warhead were far more sophisticated than previously suspected. He therefore requested that intelligence scrutiny of the Iraqi

program be made a high priority. The request was rejected and the official in question fired from his post and exiled to a bureaucratic Siberia. In explanation of this curious indifference, one former official recalls that, "We knew about their bomb program, but Saddam was our ally, and anyway, we didn't realize how far along they really were. It was off the radar." Official assessments assumed that Iraq was still ten years from producing an atomic bomb.

In fact, the bomb program proposed by Saddam back in 1979, which had commenced early operations in 1982 and gone into high gear in 1988, had been far more successful than anyone in the outside world had realized. As with any bomb program, the crucial element was the production of fissile materials—either uranium 235 or plutonium 239. Jafr and his associates pursued a variety of means for producing the requisite materials, an enormously costly approach. Simultaneously, others among the eight-thousand-strong force of scientists and technicians assigned to the nuclear weapons program toiled on a warhead design as well as a missile with which to deliver the weapon. The target date for production of a complete weapon was 1991. In fact, just before the Gulf War, the weapons design team was on the verge of success. The program to produce a sufficient quantity of uranium enriched to "bomb grade" was, however, far behind schedule. Realizing this, in the fall of 1990 the high command gave orders to take enriched uranium from the country's one officially acknowledged research reactor (which had hitherto been kept separate from the secret weapons program) and process it into bomb-grade material. Had this crash effort been concluded, Saddam would have had at least one bomb by the end of 1991.

Only in the fall of 1990, when President Bush was seeking to rally public support for war with Iraq, did Saddam's bomb become a matter for official U.S. concern. Polls showed that Americans, while generally unmoved by the fate of Kuwait and its royal family, were agreed that a nuclear-armed Saddam was a serious matter. Sites known to be associated with the Iraqi nuclear program were given a high priority in the bombing plans and were duly pulverized during the air offensive—including the plant where the "crash program" was being frantically implemented. By the end of the war, the White House and the Pentagon congratulated themselves on having destroyed the bulk of Saddam's nuclear weapons manufacturing potential. In fact, though the bombing had inflicted severe damage, the U.S. high command was being overly optimistic. The Iraqi Los Alamos, an enormous complex at al-Atheer, south of Baghdad, that was the center of the entire nuclear effort, escaped unscathed, its very existence unknown to the Americans.

Nuclear weapons were not the only "unconventional" weapons espoused by Saddam in the years before the Gulf War. Flush with his oil billions and the financial support of the Arab oil states, Saddam had embarked on a wild spending spree. Just as he gave his nuclear scientists carte blanche to investigate every possible route to the production of fissile materials for a bomb, so he authorized limitless budgets for research and development on advanced unconventional weapons of types that had hitherto been considered affordable only by superpowers. Some of these initiatives verged on the bizarre. A Canadian, Dr. Gerald Bull, convinced Saddam to invest huge sums in his "supergun," a giant cannon. Of these, the only ones that ever saw use were chemical weapons.

The British had used poison gas on fractious Iraqi Kurds in the 1920s and the Baghdad regime had used the same means on Kurds in the 1960s. In 1984, in the middle of the war he had so rashly launched against Iran, Saddam turned to this traditional Iraqi standby and began using chemical weapons in large quantities on the front lines. It proved wonderfully efficacious as a defense against the "human wave" attacks of teenage volunteers launched by the Iranians. With the help of foreign experts and a host of willing suppliers, particularly in Germany, the Iraqi chemical weapons industry made rapid progress. From the early use of mustard gas, initially developed in World War I, the local technicians and their foreign helpers rapidly progressed to "nerve agents" such as sarin and tabun, developed but never used by the Germans in World War II. Whereas old-fashioned variants such as mustard gas had to be breathed to kill the victim, the nerve gases could be deadly from the merest contact with a victim's skin.

Iraqi progress with these nerve-gas weapons was widely advertised. However, by the end of the war, Baghdad's scientists—again with German assistance—had also made significant strides in producing VX, a yet deadlier nerve agent that had the additional advantage of being safer to manufacture.

In the final weeks of the war, Saddam was preparing to put chemical warheads on long-range missiles and launch them at Iranian cities. General al-Samarrai tells of a tactical problem facing the Iraqi army, which illustrates their chilling determination to maximize civilian casualties. Staff officers were concerned that the gas, being heavier than air, would not penetrate into houses and offices in Tehran and other cities. Even those close to a missile strike might survive if they kept their windows shut. The plan devised by the Iraqi military therefore was to first send in Iraqi fighter-bombers to strike at Tehran. "They planned to bombard the city with bombs that would break all the glass in the windows," recalls Saddam's former military intelligence chief. "This would allow the gas to spread."

At least one source present in Tehran at the time believes that the Iranians were aware of the Iraqi plan and that it was one of the factors that finally persuaded Ayatollah Khomeini to stop fighting.

Even if the Iranians remained unaware of how near Saddam had come to wiping out a large slice of their urban civilian population, the Iraqi leader felt that his initiative in developing these weapons of mass destruction had yielded ample dividends. Not only had they broken the back of the Iranian mass attacks, but they had also enabled Saddam, at long last, to cow the perennially rebellious Kurds into submission. The horror of the chemical attack on the city of Halabja, in which five thousand Kurdish men, women, and children had been killed in the space of half an hour, was followed up with further attacks on civilians in other parts of Kurdistan.

According to General al-Samarrai, if the Iranians had fought on they would have been subjected to attacks from the third of Saddam's unconventional weapons initiatives. Biological weapons had been developed during World War II, principally by the British. Throughout the 1950s and 1960s, Britain, the United States, and the Soviet Union had spent many billions of dollars in refining the means of infecting an enemy with disease. But by the end of the 1960s, the international community had forsworn the research and development of such weapons, a fact that did not deter Saddam from embarking on his own ambitious program in the 1980s. The principal "agent" on which his researchers, working in deepest secrecy, focused their attentions was anthrax, a bacterium that naturally infects cattle and other livestock. When humans become infected by breathing anthrax spores, they initially exhibit the symptoms of flu and then, after two to five days, lapse into toxic shock and death.

At the time of the Gulf War, the outside world had only the vaguest inkling of the scope and success of Saddam's biological program, as it did of his nuclear efforts. Thanks to the well-advertised deployment of chemical weapons against Iran and the Kurds, the allied coalition were more conscious of Iraq's potential in that area.

In the event, Saddam never dared use chemical weapons against the allies during the war, possibly out of fear of U.S. retaliation in kind. Just before the war, Bush wrote a stern letter to Saddam Hussein demanding that he withdraw from Kuwait without conditions. The president added that "the United States will not tolerate the use of chemical or biological weapons or the destruction of Kuwait's oil fields and installations. Further, you will be held directly responsible for terrorist actions against members of the coalition. The American people will demand the strongest possible response. You and your country will pay a terrible price if you order unconscionable acts of this sort." It has been subsequently assumed that Bush was threatening to use nuclear weapons if Saddam carried out any of the "unconscionable acts" listed in the president's letter. A former senior CIA official insists that Saddam knew that "if he used chemicals, we'd nuke him." General al-Samarrai, intimate with Saddam at this time, believes that "Baghdad would have been a nuclear target." However, it appears that, in fact, the U.S. military planned to retaliate with their own chemical arsenal in the event of an Iraqi chemical attack. General Walter Boomer, the commander of the U.S. Marines in the Gulf, stated privately in September 1990 that the United States had shipped large stocks of these munitions to the region, ready for use in response to any Iraqi poison-gas attack. Although the Pentagon issued a vehement denial at the time, Boomer was certainly in a position to know the truth.

Along with his "weapons of mass destruction" Saddam had also invested large resources in long-range missiles with which to deliver them on the enemy. Not only had he acquired a large number of

Scud medium-range missiles from the Russians, his own rocket scientists had successfully labored to produce the "al-Hussein," an adapted Scud with a longer range. Used against Saudi Arabia and Israel during the Gulf War, though armed only with comparatively innocuous conventional high-explosive warheads, the missiles produced mass panic in Tel Aviv, an unpleasant reminder of what might have happened if Saddam had armed the weapons with more fearsome munitions—some of which were already loaded in missile warheads.

Thus it was that with the victory on the battlefield, the U.S. was determined that Saddam would never again be able to threaten anyone with mass destruction from chemical, biological, or nuclear weapons. The military were issuing glowing reports on the success of their bombing campaign against targets associated with these weapons programs, but just to make sure, the U.S. insisted that the cease-fire resolution passed by the UN Security Council on April 3 should provide for continued economic sanctions until there had been a full accounting of all of Iraq's unconventional arsenal. (As we have seen, Washington was determined to continue sanctions in any case.) By explicitly linking the two issues—sanctions and weapons—the Security Council was, in effect, ceding control of United Nations policy toward Iraq to whoever was adjudicating the question of Iraq's weapons. It was a linkage that was to have profound effects in the years to come.

Phrased in the uninspiring legalese of all such documents, Security Council Resolution 687 held momentous implications for the future of Iraq. Apart from stringent injunctions regarding the payment of the defeated country's foreign debts and compensation for damage inflicted on Kuwait, the Security Council ordered the creation of a Special Commission "which shall carry out immediate on-site inspections of Iraq's biological, chemical and missile capabilities, based on Iraq's declarations and the designation of any additional locations by the Special Commission itself."

Iraq was forever enjoined from developing or possessing such weapons, along with any "nuclear weapons or nuclear-weapons-usable material." According to the somewhat mangled prose of paragraph twenty-two of the resolution, once it was agreed that Iraq had complied with all the requirements on its weapons of mass destruction, the sanctions on exports of oil from Iraq would be lifted. (Paragraph twenty-one, on the other hand, made exports *to* Iraq conditional on the "policies and practices of the Government of Iraq," a far vaguer concept.) The resolution was written within a month of the cease-fire, a period when Washington still held out hope that Saddam would fall victim to an internal military coup. Thus the stipulations on the accounting for Iraq's weapons of mass destruction were written partly in the expectation that they would be implemented by Saddam's successors.

The creation of the Special Commission, with the right of intrusive inspection in Iraq, was an extraordinary imposition on Saddam, a leader who had so successfully guarded the unwholesome secrets of his regime. Under the lash of defeat, he was being told to play host to foreign "inspectors" while they conducted a fully fledged espionage operation in his own country.

The language of the resolution makes it clear that all concerned believed the task would not take very long. Iraq was given just fifteen days to hand over all information on the location, amounts, and types of its nuclear, chemical, biological, and missile programs. The Special Commission would have 120 days to develop a plan for ensuring that Iraq had complied with the draconian stipulations of the resolution. After all, given the well-advertised precision of the American bombing campaign, it appeared that this commission, made up of experts drawn from the United States, Great Britain, and other countries, would be little more than a bookkeeping exercise. This view was certainly shared by two very different men. The first was Saddam Hussein.

The Iraqi leader was well aware that the bombing campaign against his weapons of mass destruction had been largely ineffectual. A significant portion of his production facilities for enriching uranium to bomb grade had not even been targeted. The Americans were completely unaware that Jafr's engineers had adapted and advanced a method of enriching uranium through the "calutron" system, or that Iraqi missile designers were in the process of developing a homegrown missile with a range of 1,250 miles. The bombers had failed, despite enormous efforts, to hit a single mobile

Scud missile launcher during the war. Saddam still had large stocks of chemical ammunition on hand. His main biological-weapons production center at al-Hakam had remained untouched. The enemy did not even know it existed.

Thus it was that as soon as the Security Council passed the resolution, Saddam ordered his diplomats to offer full cooperation with the United Nations team that would be coming to certify that the weapons had been or would be destroyed. At a private meeting in the presidential palace, he laid out a very different agenda.

"The Special Commission is a temporary measure," said the Iraqi leader. "We will fool them and we will bribe them and the matter will be over in a few months." It was a bad miscalculation.

Among those summoned to this secret conference was General Wafiq al-Samarrai. He knew exactly what his master was talking about. Saddam "thinks that everything is possible if you have enough money," explained al-Samarrai years later. "We in the Iraqi intelligence apparatus gave presents of hard currency and gold to other intelligence agencies in the world and to officials who are now ministers in different governments." The Iraqi leader was under no illusions that American and British inspectors could be corrupted in this way, which was why he remarked at the meeting that "the inspectors should come from weak countries and from countries that believe the sanctions should be lifted."

Obviously, even the most venal inspection team could not be completely neutralized, so Saddam formulated a policy of calculated concessions. The inspectors would be given reasonably full—though not complete—information on Iraq's stocks of chemicals and imported missiles because it was assumed that the Americans and their allies would already have a great deal of data on these high-profile programs. The nuclear and biological programs would, however, be kept carefully concealed.

When the Iraqis heard who had been selected to head this "Special Commission," it seemed that Saddam's optimism had been justified. Tall and thin, with gray hair and a deceptively mild manner, Rolf Ekeus was a diplomat from traditionally neutral and dovish Sweden who had spent much of his career in the arcane world of arms-control negotiations. Iraqi diplomats remembered him fondly from 1976, when as Swedish liaison with the Palestine Liberation Organization he had worked closely with them at the United Nations in New York to get the PLO the right to speak in front of the Security Council, an effort that roused the venomous ire of the United States and Israel. In 1988, he had been his country's representative at the Conference on Disarmament in Geneva, an ongoing international negotiation crafting a worldwide ban on chemical weapons. If the Baghdad government had scrutinized his record closely at the time of his appointment to the Special Commission, however, they might have noticed an episode that gave warning of trouble ahead.

As noted, when Iraqi warplanes showered sarin, tabun, and mustard gas on the inhabitants of Halabja in March 1988, the world's governments stayed mute. No one, including the government of Sweden, wished to discommode Saddam Hussein, the hammer of the ayatollahs. Ekeus found this outrageous and informed his foreign minister that, whatever the policy, he was going to make a speech to the conference denouncing this act of barbarism—which he duly did. He was the only official representative of any government in the world, apart from the Iranians, to do so.

Like Saddam, Ekeus thought that the United Nations Special Commission on Iraq, soon to be known as Unscom, with headquarters in New York, would be a short-term operation. "I thought it should be over quickly," he said later. At the time, he also hoped it would be cheap. When he arrived in New York to take up his post, he discovered that no funds had been allocated for the fledgling organization. The only way he could raise some money was to personally vouch for a loan from the secretary general's ready cash fund, which he did with some trepidation, reflecting gloomily on the fact that he had a wife and six children to support.

A humane man, Ekeus had noted the reports of the suffering caused by sanctions on oil exports—sanctions that would remain in force unless and until he certified to the Security Council that Saddam had complied with Resolution 687. "I was in Vienna when I was appointed. I had things to do before I moved to New York, family things. But I felt I couldn't afford to wait a day," he recalled seven years later. "Iraqi oil exports had been thirteen billion dollars a year, just about thirty-five million a day. My conscience would not permit me to delay even one day. I thought, 'That day will cost the Iraqi children thirty-five million dollars.'"

Meanwhile, of course, at meetings behind closed doors at the White House, it was being decided that sanctions would stay in place as long as Saddam Hussein remained in power—that is to say, as long as he lived. At the CIA, news of the courtly Swede's appointment was greeted with a certain apprehension. "We were very, very skeptical of what began as Ekeus's open-minded approach," a senior CIA official admitted later. Nor were hard-line officials in the U.S. national security establishment thrilled to hear that Bob Gallucci, a State Department officer suspected of "liberal" attitudes toward arms control, had been appointed as the Swede's deputy.

Ekeus was determined that he should have his own people on his team—"People I trusted"—rather than a group selected by others, who would owe their first loyalties to their own governments rather than to him. He therefore set out on a recruiting drive among people he knew from his days negotiating chemical disarmament. Among them was Nikita Smidovich, an expert on chemical and biological weapons at the Soviet foreign ministry, son of a diplomat and grandson of the general who had liberated Vienna at the end of World War II. Smidovich, a burly young man with drooping mustaches, had already enjoyed the unique experience of ferreting out a forbidden weapons program—one that was being concealed from his own government.

Back in 1972, the Soviet Union, along with the bulk of the international community, had acceded to President Nixon's suggestion that biological weapons be forsaken. But, under the justification that the United States could not really have closed down their program, the Soviet military had carried on with and even expanded research and development of biological weapons. Even as the cold war wound down in the era of rapprochement between the United States and the reformist regime of Mikhail Gorbachev, this huge effort continued, employing thousands of scientists hidden away in remote institutes whose very existence was kept secret even from Gorbachev himself.

In 1989, U.S. secretary of state James Baker, apprised by his own intelligence of the secret Soviet effort, casually pointed out a building visible from the highway near Moscow down which he was driving with the then Soviet foreign minister, Edvard Schevardnadze. "That's part of your biological weapons program," he said. This was news to Schevardnadze, who kept a straight face but reported the exchange to Gorbachev at the first opportunity. Gorbachev thereupon confronted his generals, who brazenly denied the existence of the program. Schevardnadze then charged Smidovich with the task of finding out the truth. Relentlessly picking his way through the thicket of lies, half-truths, and evasions with which the Soviet military surrounded their bacteriological arsenal, Smidovich eventually forced the military to admit what was going on. It was an experience that was to stand him in good stead when Ekeus put him to work investigating the secrets of the Soviets' former ally Iraq.

Others recruited for Ekeus's commission were no less eclectic. Scott Ritter, for example, had been a career officer with the U.S. Marines. During the Gulf War, Major Ritter, then attached to military intelligence, wrote a report on the enormous allied effort to seek out and destroy the mobile Scud missile launchers deployed against Israel and Saudi Arabia. His conclusion, that not a single such launcher had been destroyed, was not only accurate but sharply at variance with the official line from the high command. Such independent thinking was not likely to enhance his career prospects. Subsequently, he fell in love with a Georgian woman and, in defiance of official edicts against such liaisons between intelligence personnel and citizens of the former Soviet Union, married her. While he was compelled to leave the military as a potential security risk, his wife was recruited by the U.S. government as a translator and duly laden with security clearances. Reviewing Ritter's résumé, Ekeus had no hesitation in signing him up for his team. They were all entering uncharted territory. Not since an interallied commission had roamed Germany after World War I in a (failed) effort to destroy the defeated country's weapons-making potential had anything like this been attempted.

Ekeus's first forays to Saddam's capital were beguiling. He was met by old acquaintances from the Iraqi foreign ministry, all speaking perfect English and observing the niceties of diplomatic protocol. It was not until late June 1991 that he had a glimpse of the real face of the regime. The occasion was the first confrontation between the UN operation and the government. The inspectors had stumbled across a crucial part of the secret nuclear program.

Technically, responsibility for dealing with Iraqi nuclear issues belonged to the International Atomic Energy Agency (the IAEA). This was the body that had regularly reported before the war that Iraq was in full compliance with the Nuclear Nonproliferation Treaty and gave no signs of a covert nuclear weapons program. It was therefore in the bureaucratic interests of the IAEA to downplay any evidence to the contrary. However, in Iraq, the nuclear overseers operated as part of Ekeus's team. Their chief was an ebullient American named David Kay.

In late June, Kay was preparing to lead a team to look at a suspected nuclear facility in a place called Tarmiya, a few hours' drive outside Baghdad. Saddam's plan to keep the scope of his nuclear effort—and the degree to which it had escaped destruction in the bombing—secret was already falling apart. A month earlier, one of Jafr's scientists had managed to make his way to Kurdistan and make contact with the American forces there before they withdrew. The Iraqi nuclear overseers were as yet unaware of the betrayal, since the scientist had faked a car accident, complete with an incinerated body. Now the defector was telling all that he knew to the Americans, information that was passed on to the inspection team setting up shop in Baghdad. Among his more interesting items of news was that Iraq had indeed found an efficient means of using calutrons—giant magnets about twenty-five feet around—tried and discarded in the United States years before.

By this time, the CIA was willing to pass on information gleaned from the intelligence satellites that hovered over Iraq. An analyst had noted that large, round objects were being moved from a heavily bombed site called al-Tuwaitha to a military encampment in the West Baghdad suburb of Abu Ghraib. (Transmitting the information was proving a more difficult undertaking. The UN team had as yet no secure communications in Baghdad, so all confidential messages had to be sent via a laborious book code, using a biography of George Bush.)

Kay and his men, crowded into two Land Rovers and a bus, took good care to arrive at the site unexpectedly, having led their Iraqi "minders" to believe that they were headed to another destination. At the gate, they were met by an astonished and angry base commander who refused to allow them to enter. Kay played the ugly American, threatening to call the UN Security Council in New York on his satellite phone. Finally three inspectors were allowed to climb a water tower just inside the fence. A few seconds later, the men on the tower radioed to Kay.

"These guys are going out the back."

One of the Land Rovers roared off in pursuit. Curiously, in view of the importance vested in the whole inspection effort by the United Nations and Washington, the men and women on the ground were operating on a shoestring. The team was using their own cameras and walkie-talkies picked up at Radio Shack. Their vehicles were castoffs from the British army, with the driving wheel on the wrong side for Iraq. The one racing to the rear of the base had a broken fuel gauge. Two miles around the fence line, it spluttered to a halt, out of gas. The second Land Rover moved out, collected the stranded inspectors, and chased the convoy of huge transporters, heavily laden with calutrons, hurriedly exiting the base. One inspector, Rich Lally, made two passes down the convoy and began to photograph the scene. Iraqi troops on the convoy opened fire over his head. He ducked under the seat and reemerged photographing as fast as he could. By the time officials cut off the Land Rover and demanded the camera and film, the roll of film was stashed on Lally's body. He refused to turn over the camera. "The last thing my wife told me before I left," Lally told Kay back at the base gate, "was, 'Don't lose the bloody camera.'"

Kay finally called Ekeus on his satellite phone and told him his men were under fire. Ekeus told him to withdraw.

The fracas at the nuclear site was a turning point. It shattered any notions that the Iraqi nuclear program had been destroyed in the war or that the Iraqis were going to cooperate with the commission. Ekeus hurried to Baghdad. It was now clear, he told Tariq Aziz at a meeting in Aziz's spacious villa near the Tigris, that "Iraq has a nuclear development program." Ever the diplomat, Ekeus chose the word "development" rather than "weapons" in order to be more polite. Unfazed, Aziz denied everything without the slightest trace of embarrassment. For the first time, Ekeus heard an explanation with which he was to become familiar over the next few years. "Do you really think we are capable of such an undertaking?" said the Iraqi with a beguiling show of sincere humility. "We are not that advanced, you know." The foreign minister and his deputy, sitting at Aziz's side, concurred, throwing in their own reasoned explanations as to why it was simply impossible for Iraq to have embarked on a nuclear weapons program.

The Iraqis were in the middle of explaining that President Saddam Hussein had ordered that the Foreign Ministry be in charge of all dealings with the UN inspectors when there was a sudden interruption as the door opened and a uniformed newcomer strode noisily into the room and threw himself on a sofa "like a spoiled child," as Ekeus later recalled. This was Hussein Kamel, minister of defense, founder of the Republican Guards, and overall supervisor of the huge effort to produce weapons of mass destruction. From the moment of his arrival in the room, he clearly dominated the group of senior Iraqi officials, who now sat stiffly and nervously in their chairs. Alone of those present, Kamel did not speak English and the translation of Ekeus's carefully enunciated points into Arabic was greeted with "coarse laughter from the sofa."

Of those in the room, only Kamel and Aziz knew that four days before, on June 30, Saddam had set up a special high-level committee, chaired by Aziz, to plan the concealment of weapons, materials, and plans from Ekeus and his inspectors.

On July 7, the committee came to a tough decision. In view of the discovery of the calutrons, Iraq would have to own up to the nuclear program. At the same time, however, the high-level group decided on a program to destroy, in secret, much of the forbidden weapons and materials on hand, the better to hide the essential items that Iraq wanted to keep. The destruction was carried out later that month at a dry riverbed close to Tikrit. This decision was to dominate the whole issue of Saddam's hidden weapons for years to come because, when Unscom deduced what had happened, the Iraqis were faced with demands to prove precisely what, and how much, had been destroyed.

While the destruction was going on, some of the more precious elements in the Iraqi weapons arsenal were being hidden away.

Some time in July, the secluded garden of a villa in Abu Ghraib was torn apart as soldiers hurriedly buried a carefully chosen selection of parts and production tools from Project 1728, Iraq's program to manufacture its own homegrown missile. Unscom would not even learn of the existence of the project for another four years. The burial party was from the Special Republican Guard, an elite unit whose duties had hitherto consisted of safeguarding the person of Saddam Hussein. The villa belonged to one of their officers, Major Izz al-Din al-Majid, Hussein Kamel's cousin and brother-in-law.

The same scene was being repeated elsewhere at other, carefully chosen sites around the country by members of this and other especially trustworthy organizations, including the Mukhabarat intelligence service and the Amn al-Khass special security service, which operated under the direct control of the presidential palace. So important was this responsibility that the small number of men selected for the task were chosen only after the most careful vetting of their families, their tribal ties, their absolute loyalty to Saddam. They were being entrusted with a secret that their leader considered almost as important as his own personal security. It may indeed be that, for Saddam, the two were one and the same. These weapons had, after all, been conceived as the ultimate deterrent against his enemies in the world outside (not to mention inside Iraq's borders, as the Kurds had discovered to their cost).

By midsummer 1991, it was becoming clear to almost everyone involved (apart from Hans Blix, head of the IAEA, who was reluctant to concede the degree to which the Iraqis had fooled him over their nuclear weapons project) that Iraq was not prepared to make the full accounting of its various forbidden weapons systems demanded, through the United Nations, by the victorious allies. Saddam had therefore embarked on an enormous gamble, since by obstructing the inspectors, he was handing Washington a ready-made excuse for continuing sanctions. As we have seen, both President Bush and his senior officials had stated, for the record, that they wished sanctions to continue until Saddam was gone. These sanctions were taking a terrible and visible toll on the people of Iraq. There was, therefore, a slim chance that outraged international public opinion might pressure the United States to lift the siege—but not while Washington and London could piously insist that Saddam was clearly defying the United Nations by concealing his weapons.

The Iraqi leader's obduracy appears entirely illogical. Yet from his point of view, the risks were evidently acceptable, especially as he initially believed he could "fool" the investigators and get rid of them in a short space of time. Back at CIA headquarters in Langley, Virginia, senior officials pondering their opponent's actions concluded that, given the ruin inflicted on his conventional forces during the Gulf War, Saddam felt he had little alternative. "In 1991, Saddam had zero conventional capability," explains one such official. "I think that right after the war he was too busy dealing with day-to-day issues to take a long-term view. He thought, 'We lost the battle, so we're going to retain and grow the weapons programs.' Even though he was constrained in his unconventional capability, that was his only option. We always thought that was what he was going to do."

"Unconventional" weapons held a natural attraction for Saddam because, after all, they had proved outstandingly useful in the past. As noted above, the chemical weapons unleashed on the Iranians had turned the tide on the front lines and had played a key role in the final and successful offensives of 1988. The threat of using chemicals on Iranian cities may also have played a role in convincing the Tehran government to throw in the towel that year. The gas attacks on the Kurds in this same period had broken the back of the Kurdish resistance and reduced the population to a state of abject terror.

Back in the 1980s, in a relaxed meeting with a group of visiting journalists from the Gulf states, Saddam related an illuminating anecdote. "When I was a child, a man walked through my village without carrying a weapon. An old man came up to him and said, 'Why are you asking for trouble?' He said, 'What do you mean?' The old man replied, 'By walking without a weapon, you are asking for people to attack you. Carry a weapon so that no blood will be spilled!'"

Apart from throwing an instructive light on daily life in Ouija, Saddam's village, the story indicates his belief in the dangers of proceeding unarmed and the benefits of possessing a powerful deterrent. He would fight long and hard to keep some vestige of his "strategic" arsenal. The United States and its allies were no less determined to deprive him of any such capability so that King Fahd and their other interests in the region could sleep in peace. Thus the stage was set for an ongoing confrontation.

One veteran CIA covert operator admits that he and his colleagues were surprised as the extent of the secret Iraqi programs slowly became apparent. "As we greatly underestimated what they were before the war, so we greatly underestimated what they were after the war. However, unlike Ekeus, we weren't disappointed to find out the Iraqis were lying."

Over the summer of 1991, the CIA began to feel they had misjudged Ekeus. Despite his unpromising record in advocating disarmament, the Swede appeared to be in earnest about probing Saddam's secrets. As a result, the agency became more willing to share information. For the first time, Ekeus was shown the actual overhead surveillance photographs and, even more important, he was told of the existence and location of a "mother lode" of documents relating to the Iraqi nuclear program. In August 1991, a carefully selected team of forty-five UN inspectors began training at a secret location in England for their most important operation yet.

The preparations were elaborate. The documents, according to the American intelligence tip-off, were being kept at the Central Records Office, in the heart of Baghdad. Ekeus and his people knew that they would not have the luxury of browsing at leisure through the vast amount of data in the building. Once the Iraqis realized what they were after, there was bound to be a confrontation. The team, therefore, trained hard on precisely which office on which floor of the building they would head for. So elaborate were the preparations for this mission that the inspectors' training site in England contained a specially constructed full-scale mockup of the Baghdad records office. They had to learn how to recognize complex documents in Arabic, how to take pictures in a hurry, and how to do all this properly the first time. There would not be a second chance.

On September 24, the team boarded a bus at the Palestine Hotel, where they lodged while in Baghdad, and headed for their target. The exercise was carefully planned to appear as a routine search in the course of which the inspectors would "discover" the nuclear weapons data. Before setting off, however, David Kay—who was once again leading the operation—suggested to U.S. network

TV correspondents in Baghdad that they should hold themselves ready to film an interesting scene that would soon unfold in the parking lot of the records building.

The journalists were not disappointed. Five hours later, the UN team reappeared in the parking lot. As they reboarded their bus, they were surrounded by armed and angry Iraqi soldiers. Inside, they had raced to the specific offices and begun copying documents, videotaping and taking pictures as they did so. When frantic security officials stopped the photocopying, the confrontation began.

The Iraqis refused to allow the team to leave without their handing back the documents already copied. Sequestered on their air-conditioned bus, the team refused to leave without them and the impasse continued into the night. The confrontation escalated as Kay broadcast his plight to the world. The evidence uncovered in the rapid search, he announced on a satellite phone from his vehicle, included the complete administrative structure of Iraq's nuclear weapons program. "We are not willing to turn over film and videotape," he declared. "This is absolutely essential for an inspection effort."

Tariq Aziz responded by accusing the UN squad of provoking the confrontation. Kay, he said, was a spy, and the seized documents were confidential personnel records. Other Iraqi officials declared that the list of names of Iraqi nuclear scientists seized by the inspectors could be used by Mossad, the Israeli secret service, to target the men for assassination. A hundred and fifty people, mostly children, who claimed to be relatives of those mentioned in the documents, duly turned up in buses armed with neatly printed banners and placards in English and French. They read: "Don't Trace Our Husbands," "Return Personel [sic] Records," and "Mossad Wants the Records."

While the operation was generally going well, Ekeus was in a fury with his deputy, Bob Gallucci. Gallucci had gone to Baghdad for the operation. Blurring the lines between his role as an official of a UN organization and his career loyalties as a U.S. Foreign Service officer, Gallucci was occupying his time by calling the State Department in Washington, whence his views were disseminated by department spokesmen. Thus, the American was implicitly confirming Iraqi government claims that the Special Commission was little more than a thinly disguised U.S. espionage operation. Ekeus ordered him to stop. Irritated officials in Washington, unable to comprehend the problem, speedily leaked news of Ekeus's order, thereby generating fresh headlines: "Ekeus Reprimands Teams." The Bush administration, paralleling Iraqi assessments, may have come to think of the special commission on Iraq as a useful arm of U.S. policy. Ekeus had forcefully to remind them that he worked for the Security Council and that Gallucci worked for him.

Meanwhile the confrontation was escalating. President Bush declared that this was a "serious business" as the Pentagon began moving troops back to the Middle East. It appeared that military action against Iraq in support of the inspectors was imminent. Perhaps because it was only just over six months since the United States had last bombed Iraq, the Iraqi government caved in on the fourth day of the parking lot siege and the inspectors were allowed to leave with their precious documents. When examined at leisure, they showed beyond a doubt that Iraq had indeed been building a nuclear weapon that could be delivered on a missile. A pattern had been established that was to endure for the next seven years: Iraqi denials, followed by partial disclosure, followed by further investigation by the UN inspectors, leading to further Iraqi admissions. Thus the Iraqi government initially denied that it had been engaged in work on a prohibited weapon—in this case, nuclear. Following the initial investigation by the UN sleuths, Baghdad admitted that it had been working on methods of producing bomb-grade fissile material—but not to the point of developing an actual weapon. After the parking lot siege, Baghdad came clean to the extent of conceding that it had been working on the "feasibility" of building a nuclear weapon.

So it followed with the other categories of proscribed weapons. Iraq conceded information on missiles imported from Russia, but long concealed "Project 1728," the effort to produce its own long-range missile. In 1991, Iraq denied ever having investigated biological weapons at all. This denial was progressively amended through no less than six "Full, Final and Complete Declarations" submitted by the Iraqi government on the subject, to the point where the Iraqis conceded the production of weapons, but claimed that all work in the area had ended in 1991 and all weapons and materials destroyed—a claim that was greeted with justifiable derision by Unscom. Iraq had embarked on a program to produce the "Al-Qa-Qa" radiological bomb, an attempt to spread radiation over a wide area. Once again, this was initially denied and then confirmed. Iraq at first denied that its chemical weapons establishment had done any research on the VX nerve agent. This was ultimately replaced with an admission that there had been such a program, discontinued because it was unsuccessful after only some six hundred pounds of the substance had been produced. Later the amount admittedly produced went up to four tons.

In 1992, the Iraqis began to explain discrepancies between the records they produced and documentary evidence cited by Unscom by claiming that the previous July they had unilaterally destroyed a huge quantity of weapons and equipment—though the original Security Council resolution had forbidden them to do so without Unscom supervision. Thereafter, the Iraqis, when pressed on weapons, manufacturing equipment, or raw materials that could not be accounted for, would fall back on the argument that they had been destroyed. The flaw in their case lay in the fact that they could never produce satisfactory documentation on the destruction. In a ruthlessly authoritarian regime such as Iraq, it was hardly credible that any officer or official would destroy important items of war-making equipment without clear, written orders from above. Yet such orders were either never produced or, when accounts of the destruction of specific amounts were produced, they turned out to be clearly inconsistent with other evidence gathered by Unscom.

The evidence emerged only slowly, an arduous process in which the scientists on the visiting Unscom mission would arrive for their inspections from their offshore headquarters in Bahrain—fully equipped by British intelligence as a secure communications center known as "Gateway"—and descend on suspect factories, offices, research centers, or anywhere else evidence might possibly be concealed. The evidence collected would then be analyzed and collated with data on Iraq's overseas purchases, reports from defectors, satellite photographs, and pictures from the high-flying U-2 surveillance aircraft lent by the United States to Unscom. Finally, Ekeus would arrive in Baghdad for one of his periodic visits and sit down across the table from the high-level Iraqi team detailed to negotiate with him and painstakingly address all the discrepancies uncovered by his team between what the Iraqis claimed and what his team had uncovered in their researches.

The process, which had brought Ekeus hurrying to New York without bothering to pack in the distant days of April 1991, stretched on into months and then years. Later, the executive chairman of the Special Commission ruefully remarked that with the exception of his wife, he had spent more of his life with Tariq Aziz than anyone else. Ekeus was under no illusion that he was dealing with fools. Such men as Tariq Aziz, Iraqi UN ambassador Nizar Hamdoon, and Deputy Foreign Minister Ryadh al-Qaysi were, as he recalls, "Very, very good diplomats. They could continue to argue for hours and hours, never forgetting a point, never getting tired." No less impressive were the technocrats like General Amer Rashid, deputy to Hussein Kamel, an engineer trained in Birmingham, England.

There was a difference, however, between the cosmopolitans like Aziz and Hamdoon, adept at detecting the most obscure nuance in a diplomatic communication, and the parochialism of even the most brilliant technocrats like Rashid. In 1994, for example, Ekeus utilized his excellent connections in New York to inspire an editorial in the *New York Times* that argued for the immediate lifting of sanctions once Iraq had complied with the requirements on weapons. A few days later, the *Washington Post* rebutted the *Times,* editorializing that sanctions should never be lifted so long as Saddam Hussein remained in power. "I don't understand America," Rashid remarked to Ekeus shortly afterward. "The *New York Times* is owned by the Jews, but they are supporting us. Yet the *Washington Post* is attacking us. Who is in charge?"

In dealing with Rashid, Ekeus had a secret weapon in Smidovich. The Russian's prior experience in dealing with his own military in pursuit of their secret biological warfare program stood him in good stead in penetrating Iraqi obfuscations. The Iraqis recognized him as a formidable opponent. On one occasion, during a discussion with Rashid on the topic of undeclared missile assets, Smidovich— who had been ordered to remain mute—began to shake his head slowly from side to side, a prearranged signal to Ekeus that the Iraqi was lying. Rashid finally exploded: "I cannot speak while Nikita is shaking his head like that."

"Okay," said Ekeus. "I will tell Nikita not to shake his head."

The discussion resumed. Smidovich once again concluded that Rashid was not telling the truth. The ends of his drooping mustache slowly lifted as his lips curled in an ironic smile. This was too much for Rashid. "I cannot talk when he is smiling," shouted the Iraqi.

While Washington may have been happy that Saddam continued to provide them with a ready-made excuse for preserving sanctions, Ekeus himself was sincere in his approach. At an early date, he pointed out to the Iraqis the significance of paragraph twenty-two of the fateful Resolution 687, which made the lifting of sanctions on oil exports contingent on Iraqi compliance with the orders to account for and destroy their unconventional weapons capability. He was making the point that it was up to him, not the U.S. government, to certify that Iraq had complied with the resolution—a reminder that did not go down well in Washington.

At times, Ekeus felt able to report a little light at the end of the tunnel. In October 1994, for example, he notified the Security Council that "the commission is approaching a full understanding of those past programs" of weapons of mass destruction in Iraq. But, in reality, the Special Commission was far from a full understanding of those programs. Two months after Ekeus wrote those words, a stocky, well-dressed man appeared at the headquarters of a Kurdish rebel group. He had walked for ten days to escape the reach of Saddam Hussein. This was none other than Wafiq al-Samarrai, formerly chief of Iraqi military intelligence, the first high-ranking officer who had served during and since the Gulf War to defect. He had been in contact with opposition groups in northern Iraq for over three years, but had kept his explosive knowledge about Saddam's secret weapons programs to himself until he was ready to make the final break with Baghdad. The wily intelligence professional had realized that the warmth of a defector's reception is directly related to the novelty and value of the information he brings with him.

An Unscom official hastened to Kurdistan and interviewed the newly arrived general. Referring to documents that had crossed his desk at military intelligence headquarters, al-Samarrai informed his incredulous interviewer that Iraq had succeeded not only in manufacturing VX, the most deadly of all chemical nerve agents, but had actually loaded it into warheads, all of which came as news to Unscom (and to the CIA and British intelligence). Furthermore, al-Samarrai revealed, the Iraqi biological warfare effort was far more advanced—and intact—than hitherto suspected. He also reported that Iraq retained a number of operational missiles, together with biological and chemical warheads.

Not everyone wanted to take what al-Samarrai said at face value. His information flew in the face of what had previously been believed about the Iraqi weapons programs. Furthermore, in February 1995, Tariq Aziz offered Ekeus the outlines of an intriguing deal. If Unscom were to give Iraq a clean bill of health on missiles and chemical weapons, Aziz would offer "help" in resolving the issue of biological weapons.

The biological warfare program had been the most secret of all Saddam's weapons initiatives, originally concealed even from officials as senior as Tariq Aziz and the army chief of staff. In its early years of operation, Unscom had discovered little about Saddam's germ weapons. In fact, the organization did not even employ a single biologist until 1994. It took U.S. congressional investigators to discover that during the late 1980s, Iraq had been buying strains of anthrax and botulinum toxins from a biological supply firm in Rockville, Maryland. The firm had failed to note or at least report the mysteriously immense quantities of these deadly pathogens purportedly required by Baghdad University, the purchaser of record—which was, in reality, acting at the behest of the Iraqi military. Unscom had been suspicious of an immense facility at al-Hakam, an hour's drive southwest of Baghdad that covered almost seven square miles of desert, but the Iraqis insisted that the buildings there, despite being surrounded by barbed wire and guard posts, were devoted to the production of animal feed and pesticides. In the absence of firm evidence, the inspectors did not probe further.

It was only after Unscom hired Dr. Richard Spertzel, a graduate of the long-defunct U.S. germ warfare program, that the search for the Iraqi biological warfare program gathered speed. In response to a round-robin sent to a number of countries requesting information on

Iraqi purchases, Israel forwarded records showing that Iraq had bought no less than ten tons of "growth media," used for manufacturing germs from the original strain, from a British company. Although growth media is commonly used in hospitals and laboratories for identifying illnesses, this was an enormous quantity, enough to make thousands of weapons. Soon Spertzel's biologists found even more, as much as forty tons that had found their way to Iraq and, ultimately, to al-Hakam.

Ekeus confronted Tariq Aziz with the irrefutable facts his staff had uncovered, but his opponent was unfazed. "When we were appointing a new minister of health," he said smoothly, "we had a choice between a medical doctor and an administrator. We picked the administrator because he was considered more loyal, and he turned out to be an idiot. He ordered far more of this material than was needed."

Despite such creative excuses from the Iraqis, the evidence was becoming overwhelming. Finally, in July 1995, Dr. Rihab Taha, an intense, British-trained scientist, sat across a table from a group of inspectors and admitted, reportedly near tears, that Iraq had indeed had a germ warfare research program, though she denied that any weapons had actually been produced. In fact, the scientist had been one of the leading lights of the biological weapons program. (Ekeus had already, unwittingly, played a big role in Dr. Taha's life. In 1993, he had brought her and the redoubtable Amer Rashid to New York for a discussion on Unscom-Iraqi relations at the United Nations. Love bloomed on the East River, Rashid left his wife, and he and Dr. Taha were married shortly afterward. "I was the matchmaker for this dreadful pair," says Ekeus ruefully.)

In August, as we shall see, Unscom received conclusive proof of the flourishing state of the biological weapons program in a most dramatic manner. As a result, the Iraqis finally confessed the true function of al-Hakam as the central germ warfare production center and, some months later, had to assist Unscom in blowing it up.

That did not end the Unscom biologists' mission, however. After sifting through thousands of pages of documents and hundreds of hours of interviews, it was still impossible to account for as many as 150 bombs and warheads that had at one time been manufactured for the Iraqi air force. For the scientists recruited by Ekeus to seek out these and other unaccounted-for weapons, that remained the central issue regardless of what was happening to the vast majority of ordinary Iraqis who had never heard of, let alone seen, a biological or nuclear weapons plant.

To a certain extent, the Unscom inspectors were shielded from seeing the effect of the sanctions prolonged in their name. They lived an isolated existence during their visits to Baghdad, and when they did venture forth, it was to encounter not starving children but willful officials who all too often impeded their work or lied to them. They were deeply convinced that Saddam had plenty of resources available to relieve the deepening misery of his people. Yet time and again they found evidence that he was instead spending money on his weapons programs. In 1995, for example, Unscom detected a covert Iraqi operation to smuggle guidance systems for long-range missiles out of Russia. On one occasion, in 1995, the Iraqis arranged a demonstration for Ekeus's benefit. A group of women attempted to deposit dead babies in his arms. This ghoulish propaganda ploy made him very angry, yet in a sense it was a case of Saddam's crudity spoiling a good case. Thanks to Saddam, the date when Ekeus could confirm that Iraq was complying with its obligations under Resolution 687 was continually postponed. Thanks to sanctions, there was a growing supply of little bodies to put in coffins.

It was all a continuation of the Gulf War. The casualty list was now growing into the hundreds of thousands, many, many more than had died in the actual bombing and fighting. The siege was claiming the lives of Iraqis who had not even been born when the war had officially ended in 1991.

FIVE

"Iraqis Will
Pay the Price"

I n May 1991, Robert M. Gates, President Bush's deputy national security adviser, had officially announced that all possible sanctions would remain in place and that "Iraqis will pay the price" while Saddam Hussein remained in power. The economic blockade instituted after the invasion of Kuwait would continue.

Two months later, in July, garbage collectors in Baghdad reported a sinister change in the loads they carted to the city dump. A year before, almost one third of all household garbage in the city had consisted of food scraps. Now, after almost twelve months of war and sanctions, the scraps had entirely disappeared. Food, any food, had become too precious to throw away. Even the skins of melons were being saved and devoured. People were beginning to go hungry.

Before the days of sanctions and war, Iraqi doctors had considered obesity a national health problem and had pleaded with mothers not to overfeed their children. While Saddam Hussein had squandered tens of billions of dollars on his war with Iran and his extravagant weapons projects, he had been careful to lavish a sizable portion of the country's oil wealth on the civilian population. The World Bank classified Iraq at the same level of economic and social development as Greece. Iraqis had grown accustomed to a standard of free medical care that would have shamed many first world countries; they took the clean drinking water that came out of their taps for granted; even the poor had become used to eating chicken at least once a day.

There were few chickens left now. Uncountable numbers of these birds, bred from a strain developed by U.S. agricultural scientists that became known as the "Iraqi chicken," had died when the electricity powering their modern henhouses had abruptly stopped flowing after the first bombing attacks on the power stations. The Iraqi chicken had been dependent on a carefully designed diet imported from abroad. Sanctions prevented any further supplies of the feed crossing the border, and in the market, a single sad-looking bird cost the equivalent of thirty-seven dollars. Egg production had dropped from two billion a year—two a week for every Iraqi—to just two million.

The rich diet, the government-financed trips to London or Paris for specialized medical treatment, the clean water supplies had all been paid for by oil exports—$13 billion in 1989. Two years later, the government would garner just $400 million from oil smuggled abroad under Turkish trucks in defiance of the sanctions mandated by the United Nations. That first summer after the war, Doug Broderick, a professional aid worker shipped to Baghdad by the U.S. charity Catholic Relief Services, cast a professional eye on the breakdown in the health system, food supplies, and the overall effect of sanctions on the economy and soberly forecast to us that, as a result, no less than 175,000 Iraqi children would inevitably die.

For the poor, the effect of an overall 2,000 percent increase in food prices within a year of the Kuwait invasion was devastating. In the vast working-class suburb of Saddam City in the east of Baghdad, the streets were dotted with what looked like heaps of rubbish. On closer examination, these turned out to be bundles of torn rags that people were trying to sell as clothes.

For visitors to Iraq, the most striking evidence of the devastation of Iraqi society was the plight of the middle class.

When the hundreds of thousands of Kurds fleeing Saddam after the postwar uprising had appeared on network TV news in early April 1991, Western audiences had been moved to pity partly because so many of the refugees shivering on the bleak mountainsides looked "like us," doctors and lawyers in three-piece suits. Similarly, the sight of highly educated professional Baghdadis, who would not have appeared out of place in an American or British suburb, sinking inexorably to a subsistence level was in a way more striking than the plight of those who had always been poor.

On a blistering hot Thursday that first July after the war, a crowd of men and women surged against the locked gates of St. Fatima's Church in a quiet and prosperous-looking neighborhood near the center of Baghdad. Some of the women were dressed in the black chadors of the lower classes, but an equal or greater number showed by their Western suits and high heels that they were solidly middle class. They had come to this Christian establishment because Catholic Relief was about to distribute food. Behind the gate, Doug Broderick was explaining how Iraqis were responding in ways that were all too familiar to him given his experience of famines around the world:

"Right now throughout the country, we have a classic response to a food shortage, pre-famine. You have people selling jewelry here in Baghdad. Your used-watch market is flooded with watches. Families are pawning their carpets, their furniture, their gold, their silverware. Anything that has any kind of value—their cameras, their videos, their radios—in order to get cash for food."

There were now hundreds of shouting applicants on the other side of the wrought-iron gates waving the bits of paper that entitled them to a food handout. The temperature was approaching 120 degrees. The women in front were being crushed against the gate, their faces squashed flat. Suddenly there was a roar as the gate buckled and crashed open, the crowd rushing through.

Once inside, the rioters' raw panic that they would be turned away empty-handed subsided. The ladies straightened their skirts and examined their broken heels. The men brushed off their jackets. A line formed as if for a Safeway checkout counter. They were once again middle class and civilized.

As is common in many of the oil-rich countries, the Iraqi government is a major employer, once in a position to pay generous stipends to civil servants, doctors, and university professors who formed the backbone of the middle class. Now, with rampant inflation, their once adequate salaries were being reduced, in real terms, to a pittance.

Mohammed Jawad*, a handsome, well-built man, was a professor of engineering at Baghdad University, where he had taught for twenty-five years. He had a pleasant house with a shaded garden in an affluent suburb and he drove a late-model white Subaru. Hala, his wife, was a talented decorator. Eighteen months after the invasion of Kuwait and the disasters that followed, his salary had been reduced by inflation to the equivalent of five dollars a month. He expected that his pension would amount to seventy-five cents.

Jawad had nearly escaped. "I had just been offered a job in Germany when Saddam invaded Kuwait," he recalls sadly. "But I came back to see a student through the final stages of his Ph.D." The sanctions imposed by the United Nations on Iraq four days after the invasion of Kuwait made it impossible for ordinary Iraqis to travel. "I was trapped here and haven't been able to get out since." He wanted to resign from the university and find work elsewhere, but the Iraqi government had forbidden anybody to leave government service. In the meantime, he was able to supplement his nearly worthless salary with some consulting work for private companies rebuilding bridges and offices hit by allied bombs. In that first year after the Gulf War, he had hopes that both sanctions and Saddam Hussein would speedily come to an end. Talk against the regime among his friends was fairly open. At that time, the security services, though very active, had more serious foes to pursue than academics, whose opposition they knew would never get beyond words.

Twelve months after the war, the Jawads threw a dinner party for a group of close and trusted friends. Hala produced an exiguous casserole composed of chickpeas, yogurt, and canned American ham. The ham had been part of the relief supplies airlifted by the United States the previous year to Kurds starving on the mountains of the Turkish border after fleeing the counterattack that had followed the uprising. Being Muslims, explained the hostess, "Kurds

---

*Not his real name.

don't eat ham. Only the Christians and secular nonbelievers like us do. They dropped this on the Kurds and we get it now in the market."

The talk around the table dwelled on the catastrophes of the previous year and the repeatedly expressed belief that sanctions, in place since six months before the war, had to end soon: "It has been eighteen months now. They cannot go on much longer." Sanctions, as several of the guests noted, were reinforcing the regime's line that the war had not been about the liberation of Kuwait but had been a direct attack on Iraq and its people. Soon it was time to switch on the TV to watch *Portfolio,* a hugely popular series on the war narrated by an anchorman who modeled his style on that of Alistair Cooke. Subversive murmurs about "him" died away as the images, pirated by state TV from foreign broadcasts, of the six-week bombing campaign flickered across the screen. The audience in the Jawad living room repeated the anchorman's solemnly intoned statistics: "Four kilos of ordnance for each Iraqi."

It was one of the Jawads' last dinner parties. Not long afterward, Mohammed began meeting old friends at a *schwarma* sandwich bar in a crowded street in North Baghdad. His stated reason for no longer entertaining at home was security. The attitude of the secret police was becoming harsher. "You cannot imagine the fear in the hearts of the people," he explained. But there was a second unstated reason for why he now took guests to a cheap café. He had become too poor to dispense hospitality.

Pathetic signs of growing middle-class poverty were becoming more evident. In the Souq al-Sarrai, in a yellow-brick passageway off al-Rashid Street in the center of Baghdad, a book market had appeared where shoppers could buy Dostoyevsky or a copy of Plutarch's *Lives* in English for the equivalent of fifteen cents. Iraqi intellectuals were selling their books, sitting on the sidewalk beside a heap of old volumes. Often the flyleaf of the book showed that it had been bought in the 1930s by some eager Iraqi student in Britain, where so many of them had been educated.

The Iraqi middle classes were plunged into lives of deep insecurity. Many of them had survived the political turmoil of the previous thirty years surprisingly well. They were conscious and proud of the intellectual history of Iraq, stretching back to Babylon and Ur, but by the early 1990s, they were desperate to escape. It was an ominous development. The existence of this large, highly educated, and secular group had helped propel Iraq out of the third world, constituting a resource hardly less valuable than the oil fields, certainly contributing more to the country's underlying strength than Saddam's vainglorious military projects.

Mohammed Jawad applied to dozens of universities and colleges in the U.S. and Britain, where he had once been a welcome visitor, but now even the mechanics of sending a letter of application through Jordan was complicated. His efforts were, in any case, without success. The fear and loathing that Saddam had evoked in the West by his brutality, heavily publicized since the invasion of Kuwait, was being indiscriminately applied to his hapless people. It gradually dawned on Jawad that *every* Iraqi was regarded as a pariah by the outside world.

Other Iraqi academics were willing to take any opportunity, however humiliating, to get out. Four years after the war, Muammar Qaddafi, the Libyan leader, expelled Palestinians who had worked in Libya for decades as teachers, accountants, and low-ranking officials. His motive, never entirely clear, was apparently to add to the number of Palestinian refugees—and underline to the world that the Oslo Accord, between Israel and the Palestinians, was doing nothing to return them to their homes. It was an ineffectual gesture, but it left Libya short of teachers. A Libyan mission came to Iraq to recruit Iraqi replacements for the Palestinians. News of its presence spread among Iraqi intellectuals. Most of the jobs were at a lowly level, teaching children to read and write. But outside the Libyan embassy, in the Mansour district, there was a near riot as Iraqi academics clustered around the gate. Dressed in suits and smart dresses, they pressed against the railings, clutching documents in several languages to prove they held Ph.D.s or were multilingual. Jawad was among those who applied, but without success.

The professor had at least by that time finally freed himself from his university job. He had succeeded only through a complicated process of fraud and bribery. "I tried to resign for two years," he explained later. "All the professors were trying to leave. Eventually I bribed a doctor to say I had a serious heart condition. Even so, I had to spend two weeks in the hospital, where they gave me medicine I didn't need. They even hung fabricated charts at the end of my bed showing I was really ill." He laughed sourly at the way he ended his university career: "What a way to finish twenty-five years of service."

Iraq, which before and long after Saddam took power, had been noteworthy in the Middle East for the rigorous honesty of its civil service, was becoming a corrupt society as officials extracted meager payments, even food, in exchange for routine services. Nuha al-Radi, the leading Iraqi ceramicist, who kept a diary of the war and its aftermath, records the payment of the equivalent of five dollars and a bucket of yogurt for the renewal of the license on a car.

Raging inflation was fueled by the government's resorting to the printing press for money, but others were manufacturing currency too. One UN agency, establishing itself in Baghdad, noticed that the first piece of equipment its Iraqi staff asked be sent from Jordan to Baghdad was an expensive color photocopier. Officials suspected that it was needed to photocopy newly issued and easily forgeable twenty-five-dinar notes, which were replacing the old Swiss-made currency. Iraqi shopkeepers all started using blue lights under which they could detect the palm tree watermark on government-issue notes. In 1990, one Iraqi dinar was worth $3.20. Five years later, a single dollar bought 2,550 dinars. In Baghdad, money changers provided plastic bags in which to take away weighty bundles of notes.

Many Iraqis stopped using money at all. Nuha al-Radi recorded one person renting a room for two trays of eggs a year, but she notes that another family was so poor that it could not afford the annual rent demanded for a house: one chicken. All serious business was done in American hundred-dollar bills, and after 1996, only the new hard-to-forge issue, with a bigger picture of Benjamin Franklin on the front, was acceptable. They were known as "phantoms" because Iraqis thought Franklin's domed forehead and wispy white hair gave him a ghostly appearance. One Iraqi said: "A hundred dollars is worth so much to us these days that we can't afford to be taken in by a forgery."

With society fraying at the edges, life in Baghdad and beyond was becoming dangerous. Law and order had never been a problem, perhaps unsurprisingly. People routinely left their doors unlocked. Now conversations at gatherings like the one at the Jawad residence dwelled increasingly on anecdotes about the latest crimes and their victims. The United Nations had four vehicles stolen, mostly at gunpoint. Everybody in Baghdad had stories of daring or ingenious thefts and the ruthlessness of the thieves. In one case, a pious woman gave a cooking pot full of dolma, or stuffed vegetables, to the guards at the Abu Hanifa Mosque, the most famous Sunni Muslim mosque in one of the oldest districts of Baghdad. The food was heavily drugged. While the guards slept, reputedly for two days, thieves removed ancient carpets and even the enormous chandeliers.

There were other signs that people in Baghdad were frightened of robbers. Every Friday there was a dog market in the souq al-Ghalil, a stretch of empty ground beside the main road on the fringe of the principal market district. In the years after the Gulf War, guard dogs were in great demand. They came in all sizes, from dapper terriers to grim German shepherds. "Nimr is a clever dog," one man told us, pointing to the pugnacious hound beside him. "He will tear a piece out of any enemy who gets near your home. But if he sees somebody whom he knows is a friend of his owner, he will never attack him." Nimr was on sale for the equivalent of eighteen dollars and looked worth every cent.

Not everyone was suffering. Early in 1992, Hala Jawad found she could put her artistic talents to work designing enormous floral bouquets for the lavish wedding parties being thrown by the "new billionaires," businessmen who were profiting hugely from smuggling as well as from the lavish government contracts for reconstruction work on bombed buildings, power stations, and bridges. The wedding parties, thrown at the al-Rashid Hotel or other upmarket venues, were in lurid contrast to the fraying and increasingly desperate society outside. In the vast hotel ballroom, rows of tables groaned under plates of food, with a fifth of Johnnie Walker Black Label (the Iraqi national drink) for every couple.

The auction houses and antiques dealers provided a point of connection between the new rich and the new poor in the city. At the Baghdad Auction House, close to the city center, the weekly auctions generated a throng of once comfortably off intellectuals and professionals like Professor Jawad jostling to see what treasured family heirlooms—carpets, furniture, paintings—might fetch from the black marketeers and senior Baathists at their elbow. Also among the bidders were dealers from Jordan, pouring into Baghdad to snap up the accumulated possessions of the dying middle class.

Adding to the resentment the new poor felt toward the United States and its allies was the fact that the native new rich had been supplemented by an influx of UN officials, a new colonial class, who were highly paid in hard currency. At the heavily guarded UN headquarters in the Canal Hotel in East Baghdad, where Professor Jawad at one point thought of applying for a job as a driver, an official proudly pointed out to us the two glorious carpets on his office floor, each worth $1,500, which he bought for $40 in Basra.

Even that outrageous bargain might have made the difference between life and death for the seller. Real earnings fell by 90 percent in the first year of sanctions, and then by another 40 percent over the next five years. Monthly earnings for Iraqis employed by the government fell to five dollars a month. In the Alwiya Hospital for Obstetrics and Gynecology in Baghdad, there was no water available for washing mothers and their newborn children. Patients were told to bring their own mosquito netting. In another hospital, a team of Western doctors observing the state of the Iraqi health service "witnessed a surgeon trying to operate with scissors that were too blunt to cut the patient's skin." Hunger led to a return to ancient means of securing food. Visitors noticed that after the main Iraqi grain harvest in May, women were gleaning, walking through the fields looking for stray grains of wheat left behind.

The decline did not follow an even pattern. The sanctions introduced immediately after the invasion of Kuwait had effectively banned food exports to Iraq, as well as all other goods, with the exception of medicine. For a short period, the effects of the blockade were masked by an inflow of goods looted from Kuwait. Then, thanks to the allied bombing and the rebellion that followed, the economy basically came to a halt. There was no electricity to pump water and sewage because the power stations had been bombed, no fuel for transportation because the refineries had also been hit. The rebellions dislocated all administration in the Shiite and Kurdish areas as officials fled for their lives. Hyperinflation had set in—prices rose 600 percent in just the first six weeks after the war and continued to rise.

Given such a terrible situation, it is little wonder that in the first few months after the war, sober and qualified observers reported the situation in cataclysmic terms. The United Nations official Martti Ahtisaari, who visited Baghdad in the middle of March 1991, stated on his return that "The recent conflict has wrought near apocalyptic results. . . . Iraq has, for some time to come, been relegated to a preindustrial age." Three months later, Prince Sadruddin Aga Khan, a special envoy dispatched by the UN secretary general, toured the country and reported that "the rapidly deteriorating food supply situation has brought the Iraqi people to the brink of a severe famine" and predicted imminent "massive starvation" and spreading disease.

In the event, although disease did inexorably spread through the population, there was no immediate massive starvation complete with scenes familiar to television viewers from famines in Sudan, Ethiopia, and Somalia. Iraqis were at least able to survive thanks to the government's system of rationing. From the introduction of sanctions in 1991, all Iraqis were registered at one of fifty thousand private shops that acted as agents for the state. Here the customer could buy, for a nominal sum, seventeen pounds of wheat flour, three pounds of rice, half a pound of cooking oil, three pounds of sugar, and just over three pounds of baby milk, where required, as well as two ounces of tea. There were also allowances for soap and washing powder. The ration was reduced in 1994, but still provided about 53 percent of the minimum food needed for an adult Iraqi to stay alive. A survey of households by a team of Western specialists in 1996 "determined that the system is highly equitable and appears to be one of the most efficient distribution systems operating in the world." (One consequence of this system, perhaps little noticed in Washington, was that dependence on rations from the state actually strengthened the government's control over the people.) The crowd at the gates of St. Fatima's Church would not starve without their charity handout, they would merely be very hungry. For Westerners accustomed to confronting third world poverty only when emaciated corpses begin appearing on TV, this might not seem such a dire situation. But for a mother whose monthly ration of baby milk ran out after two weeks, it was quite terrible enough.

Under the postwar UN sanctions regulations, Iraq was allowed to import food and medicine, but these had to be paid for and, given the embargo on oil sales, it seemed hard to explain where the Iraqi government was finding the money to import food for the ration system. The first move made by President Bush in response to the invasion of Kuwait had been to freeze Iraqi financial assets in American banks, a move followed by the rest of the coalition. It was known that at least part of these assets had escaped seizure, but no one knew how much. While in exile after the invasion, the government of Kuwait hired the New York investigative firm Kroll Associates to try and find Saddam's hidden treasure. The search was not a great success, although the firm did identify some hidden Iraqi holdings overseas, including a stake in the French firm that publishes *Elle* magazine. Early in 1991, Kroll reported that Saddam might have as much as $5 billion squirreled away. The estimated $5 billion would not have been sufficient to meet the basic needs of the Iraqi people for very long, even if Saddam had been disposed to drain his entire bank balance on their behalf. In the days of affluence, Iraq's food import bill alone had been running at about $4 billion annually.

Part of the answer to the puzzle of the financing of food imports could be found across the border, in Amman. The Jordanian capital had become the entrepôt for trade with Iraq, and the merchants of the city, many of them exiled Iraqis, were doing a roaring trade. Iraqi officials were turning up around the Middle East with gold ingots for sale; scrap metal and industrial machinery were being smuggled abroad by the truckload; loot from Kuwait was also on offer, for which there was a good market in Iran. Ironically, much of the trade with Iran depended on cooperation with the rebellious Kurds, since they controlled the border crossings favored by the smugglers. There were more direct routes to Iran farther south, but the bribes extracted there by local Iraqi army commanders were considered by the Amman traders to be outrageously high: "Fifty percent! Who can make a profit on that?" The Iranians paid for the goods in dollars, which then came west to Jordan, Syria, Turkey, and beyond to buy what was needed. In addition, some adventurous speculators were paying dollars for controlling stakes in Iraqi state corporations.

A gathering in an out-of-the-way office in Amman a year after the war provided a telling insight into the way in which Saddam enlisted the free market to help feed his people. A young Baghdadi trader in a leather jacket and jeans had arrived to do business with the businessmen gathered in the office, an affluent and well-tailored group who appeared to regard him as a somewhat crass arriviste. The Baghdadi, like everyone else in the room, wore a Rolex and was in the market for luxury goods. However, he announced that his "conscience" impelled him also to buy sugar for donation to the government's ration effort.

Once he had departed, his stated motives for generosity came in for some scornful commentary. "Bullshit," said one of the locals present. "It's not his conscience that's making him buy sugar, it's his fingernails." This was an allusion to the routine administration of the "Iraqi manicure" in the regime's torture chambers, involving the extraction of fingernails and often toenails as well. In other words, the leather-jacketed trader was under duress to import a basic commodity like sugar in exchange for leave to pursue his more profitable activities. Sometimes, more than fingernails were at risk. In July 1992, forty-two merchants were summarily executed in front of their shops in the market district of Baghdad for "profiteering."

Iraq had a further resource in its own agriculture. This had been neglected during the years of the oil boom, since it seemed far easier to import rice from California or beef from Ireland or wheat from Australia. After the war, following decades in which Iraqis had flooded into the cities from the countryside, the flow was suddenly the other way. Perhaps a third of the Iraqi population lived in the greater Baghdad area. But many were recent immigrants from the countryside and still had links with their villages. Now they began to desert the big city and go home. Suddenly the fields of the Mesopotamian plain were full of people again working much as their ancestors had done before mechanization. Khalid Abdul Munam Rashid, the agriculture minister, explained: "Because of the lack of machinery, we do more things manually, using eight people where we used to use two." The Ministry of Labor and Social Welfare estimated that nearly 40 percent of the Iraqi population was engaged in agriculture in some way, three times the number before the invasion of Kuwait.

The government was careful to make sure that agriculture was worth the farmers' while, steadily paying them the equivalent of about a hundred dollars a ton for their grain despite the collapse of the local currency. That was forty dollars a ton more than they had received before the war. Big landowners made money. "They come in here and buy mirrors and chandeliers," said one antiques dealer of the newly rich agriculturists. With the traditional distaste of Baghdadis for Iraqis who live in the countryside, she added: "They have very bad taste."

As the government displayed unexpected resourcefulness in alleviating the food crisis, so too did it appear to have performed miracles in repairing the bomb damage to vital systems like the power plants. The reconstruction effort, billed as "Hujoum al Mudhad," or "the Counterattack," was a crash program initiated within weeks of the end of the war. Its effective moving spirit was a brisk British-trained technocrat named Saad al-Zubaidi. He was given the task of repairing the damage without importing parts and expertise. A year after he set to work, he rattled off cherished statistics on his achievements, wrought with a "blank check" and a ministry staff of twenty-eight thousand people—"We reinvented the suspension bridge."

As the house Zubaidi built for himself on a palm-lined street in the upscale district of Mansour attested, the rewards for the heroes of the counterattack were considerable. Anyone lucky enough to own a dump truck could receive a thousand dinars (roughly five times the average monthly wage in that period) a day. Rooting in his drinks cabinet for a bottle of Glenfiddich ("No, no, Black Label is not good enough"), Zubaidi talked of the pride he felt that Iraqis were rebuilding on their own what before they had depended on others to provide. "More than ninety percent of the major bridges were built by foreign companies. The telephone exchanges, the power stations. All oil-exporting countries had a similar disease. They were totally dependent on foreigners. It was easier to send a telex to Japan when you needed something."

Zubaidi, as a favored and richly compensated servant of the regime, might have been expected to talk like this, but pride in the reconstruction effort ran across the political spectrum. "We did it by ourselves!" exclaimed Professor Jawad when the government announced that fifty bridges had been rebuilt.

The achievements of the counterattack indeed appeared impressive. The Jumhuriya Bridge that linked the center of Baghdad across the Tigris, smashed by allied bombs, was rebuilt within a year. By May, all Iraqi provinces, known as governorates, were reconnected to the national electricity grid. The huge power station at al-Dohra in South Baghdad, which supplied the bulk of Baghdad's power, had been put back in working order by cannibalizing parts from other power plants. One of its four tall chimneys, brought down by allied bombs, had been rebuilt and painted in the Iraqi national colors. Al-Hartha, the main power station for Basra, had been reduced to a pile of tortured metal after no less than thirteen allied raids during the war. Just over a year later, it was again, miraculously, producing power.

Once the electricity was flowing again, the situation appeared to brighten. Within four months of the end of the war, Iraq was generating 40 percent of the electricity it had produced the previous year. Few factories were working, so most electrical power went to consumers. The urgency of the first, doom-laden reports from concerned foreigners had been partly inspired by the fact that the wartime shortage of gas had brought the country's food transportation system to a halt. They thought the halt was permanent. But with power, the repaired refineries (most of which had escaped total destruction because one farsighted official had ordered them drained of oil just before the bombing) could produce fuel. Gas was available again, almost for free. Now food could be trucked into the cities from the countryside and Jordan. Mohammed Jawad could once more take the Subaru to work, even if his tires were bald and almost impossible to replace.

When the power was still almost totally out, aid workers and local officials had predicted plague because sewage could not be pumped. Just after the intifada, Khalid Abdul Monem Rashid, the mayor of Baghdad, told us: "An epidemic could easily kill fifty thousand in Baghdad, if we cannot control the sewage." That threat retreated when it became possible to at least pump sewage into the river.

The government was able to halt deterioration in other areas as well, though sometimes by savage means. Three years after the war, in response to the rising crime wave, the Revolution Command Council decreed that anybody convicted of robbery or car theft would have their right hand amputated at the wrist. If they repeated the offense, they would lose their left foot. Iraqis knew the decree was to be taken seriously because on the day it was published, *Babel,* the newspaper founded and controlled by Uday, the son of Saddam Hussein, said that failure to implement it would be damaging. The paper cited the failure to implement an earlier decree on the execution of the madams of brothels as an example of unpardonable laxness. In June 1994, Baghdad radio began routinely reporting court sentences such as the amputation of the hands of two people convicted of stealing carpets from a mosque in Baquba, northeast of Baghdad, and a similar dismemberment for the woman who stole a television from a relative, sentences all handed down in one day.

Such punishments were, of course, in line with the traditional penalties of Islamic law, but it was a novel direction for Iraq. Before and after the Baathists took power, Iraq, at least in the big cities, had always been a refreshingly secular society. Visitors from the bleak environment of Saudi Arabia noted with relief that women were under no compulsion to conceal themselves under veils and scarves. Indeed, women were encouraged to pursue careers; in the first stages of reconstruction, many of the bustling building sites appeared to be under the direction of startlingly attractive lady engineers, yellow hard hats perched on top of their flowing tresses. Alcohol was freely available and consumed in staggering quantities by those who could afford it— the minimum order for a scotch in the Baghdad nightclub district near the U.S. embassy was a quarter bottle.

Within three years after the war, this easygoing atmosphere began to change. The impulse came both from below and above. As their standard of living collapsed, the salaried classes in Baghdad increasingly took refuge in religion. Attendance at prayers in the venerable mosques of the city soared. Saddam noted the trend and shrewdly moved to capitalize on it, as with his institution of amputation as a response to crime, by increasingly casting his regime as "Islamic." The government was also astute in banning the public sale of alcohol. Ordinary Iraqis were deeply resentful of the fortunes being made by black marketeers with connections to the regime. This resentment was fueled by garish and public displays of newfound wealth by the new billionaires. The lavish weddings at the al-Rashid, which generated a modest pittance for Hala Jawad with her floral bouquets, were one much-noted example. Word of other extravagances quickly circulated among the population, as in the case of one suddenly wealthy entrepreneur, moved to ecstasy by the nightclub performance of a talented belly dancer, who hurled a blank check at her feet. As the incident gained notoriety, the high command took notice. "Chemical Ali" Hassan al-Majid denounced "denigrating nocturnal activities," and the patron of dance was thrown in jail and fined the equivalent of twenty-five thousand dollars.

More general measures followed. By 1995, nightclubs and discotheques were closed. Restaurants were no longer allowed to serve alcohol, although Iraqis could still buy whisky, arak, and beer from Christian-owned shops, which were specially licensed. It is a measure of the fear with which the government was regarded that restaurant owners did not dare break the new rules, even though it sometimes meant their financial ruin. Restaurants like the al-Mudhif, a well-established Lebanese restaurant in Abu Nawwas Street, became completely dry. When a foreigner brought in a hip flask of whisky, a frightened waiter, putting his wrists together as if he were handcuffed, whispered: "Put it away. Do you want me to go to jail?"

In 1994, Saddam decided on a very concrete reaffirmation of his Islamic credentials. He announced that the largest mosque in the whole world would soon be built in Baghdad. A large site was available at Muthanna, the capital's old municipal airport that had been wrecked by allied aircraft and missiles in the war. The great structure would be known as the Grand Saddam Mosque.

This was the first big building project in the country since the end of the frenzied reconstruction phase in 1992, and it was good news for hungry specialists like Professor Jawad. Although Saddam appointed himself chief engineer, eleven design teams were set up, and there was a job for Professor Jawad on one of them. The plan was to build a concrete dome the size of a football field. It would rise from the center of an artificial lake, in the shape of the Arab world and fed by the Tigris. On entering the mosque, the worshiper would see an electronic picture of Saddam. Four massive towers at each corner of the lake would house an Islamic university.

It was an empty dream. Once upon a time, Iraq could have furnished the resources to build such a grandiose extravagance, but not now. Jawad and the other professionals knew that the country simply did not have the materials or the equipment for such a project and that they could not be imported due to sanctions. The most basic requirements were totally out of reach. "We do not have high-tensile steel, pile drivers, reinforcement bars, or additives for the cement," he said at the time. In the first year of construction, the only part of the mosque to be completed was an elegant pavilion from which the chief engineer could gaze on his barren site.

The Grand Saddam Mosque was, in a sense, a metaphor for the hollowness of the whole "counterattack." On the surface, Iraq appeared to have successfully surmounted the threat of plague and famine sincerely predicted by aid officials after the war. Power stations had been brought back to life and pools of filth on the city streets receded as sewage once again flowed through the pipes, at least in some areas. The ration system kept people from starving, and it appeared that the economy of the country might be able to sputter along at a minimal level pending the day when sanctions were finally lifted.

However, like the plans for the great mosque, the reconstruction effort had turned out to be a chimera. The power stations had been repaired with parts cannibalized from others, parts that could not be replaced when they broke down. Bombed factories might have been rebuilt, but raw materials could not be imported to supply them. New fields had been planted, but there were only scant amounts of vital pesticides, fertilizers, animal feed, and spare parts for irrigation machinery.

A glance at the Tigris indicated, both literally and figuratively, how sick the country really was. When Hulagu, the grandson of Genghis Khan, sacked Baghdad in 1258, Iraqis say the water in the river Tigris changed color twice. On the first day, it turned red with the blood of the thousands slaughtered by the Mongols; on the second, it went black because of the ink from the books—from what were then the greatest libraries in the world—which Hulagu threw into the river.

In the 1990s, the Tigris changed color again. It was now a rich café-au-lait brown because raw sewage from 3.5 million people in Baghdad, not to mention effluent from the cities upstream, was entering the river. The partly revived power system had made it possible to pump the sewage out of the sewers. Formerly it would have gone to the highly up-to-date and efficient treatment plants before finally being put in the river. But the plants were not and could not be repaired, so the sewers emptied straight into the river.

The treatment plants remained idle for various reasons, all of them man-made. They had originally ceased operation when the power stations were hit. Sewage had immediately begun backing up in the system, in some places causing the pipes under the city streets to rupture, leading, in turn, to the very noticeable pools of sewage lapping at people's doorways at that time.

Restored power at least made it possible to move sewage through the sewer system in some areas, but meanwhile, some of the treatment plants themselves—including al-Rustamiya, the main Baghdad facility—had been hit by bombs. Treatment plants are highly complex facilities and, in any case, need constant maintenance and repair. Even without the need to repair bomb damage, Iraq needed a constant supply of spare parts from abroad, but the Iraqi government could not or would not supply the hard currency needed to buy them. Even when the necessary parts were bought, they could not be imported until they had gone through time-consuming scrutiny from the sanctions committee.

There was, however, an additional problem. Sewage treatment plants rely not only on machinery but also on chemicals, the most important being chlorine. Chlorine is also used in the production of chemical weapons. As a "dual use" item, therefore, chlorine was subject to tight restriction by the sanctions committee. Imports were effectively limited to supplies brought in by UNICEF. Iraq did produce some chlorine on its own, but almost all this supply was reserved for use in the separate and even more vital plants that treated drinking water. Even so, the amount available for the drinking-water treatment plants amounted at the best of times to only 70 percent of what was needed. When the treatment system failed, engineers simply pumped untreated water through the system.

For all the euphoria of the "counterattack," therefore, the most basic requirement of clean water had not been met. Members of a Harvard School of Public Health team who reported on the breakdown of the water and sewage treatment system in late 1991 found that almost nothing had changed when they returned in 1996. "Water plants throughout Iraq are now operating at extremely limited capacity," they reported after the second trip, "and the sewage system has virtually ceased to function."

The end results were visible in the children's wards of the hospitals. To cite just one statistic, every year the number of children who died before they reached their first birthday went up, from one in thirty the year sanctions were imposed to one in eight seven years later. Health specialists agreed that contaminated water was killing the children. The dirty water that brought gastroenteritis and cholera in its train found its little victims easier to overcome because they were already weak. With malnutrition, the immune system is weakened, particularly in children. Iraqis, especially their children, were not getting enough to eat.

"An average monthly salary buys just two chickens," explained Viktor Wahlroos, the deputy coordinator for UN relief operations in Iraq, in 1995. "A quarter of the children are suffering from malnutrition. The government ration meets fifty percent of people's needs and they don't have the money to buy the other half." Begging had become commonplace in the main streets of Baghdad. Children would approach cars waiting for the traffic lights and hold on to door handles and wing mirrors when the car started to move. They would only let go when given a small sum of money. A study of 2,120 children under five years of age in Baghdad carried out in the summer of 1995 showed how far their health had deteriorated since the war. In 1995, 29 percent were underweight, compared to 7 percent in 1991. The number of children classified as "stunted" had risen from 12 percent to 28 percent. The study's authors said such conditions were comparable only to infamously poverty-stricken countries such as Mali.

By the late 1990s, it was becoming difficult to go anywhere in Iraq without seeing signs of the disintegrating infrastructure. Diyala province, east of Baghdad, is potentially rich, with soil well-watered by the Diyala River, a tributary of the Tigris, flowing from the mountains of Kurdistan. At first sight, the farmers in the village of al-Yaat on the banks of the river did not look like victims of sanctions. They had good land, grew their own food, and could take advantage of high prices for their fruit. "It looks as if we are well off," said Buha'a Hussein al-Sayef, one of the largest farmers in the village, as he sat on the balcony of his spacious house overlooking a garden shaded by date and pomegranate trees. He admitted that people in the country were better off than those in towns, but then he listed what the villagers lacked. Their small water-purification plant had long ago ceased to work. They pumped contaminated irrigation water into their homes. They also worried about their health and that of their relatives. Buha'a Hussein introduced his cousin Ahmed, a visibly ailing twenty-four-year-old, who had been operated on at the Cromwell Hospital in London in 1985 for heart problems. He was meant to have further surgery, but the family was unable to pay for it.

Because medicine was in short supply and hospitals deteriorating, Iraqis came to believe that almost any disease might be curable if it were not for the sanctions. In the Iraqi countryside, villagers would often keep dusty X rays of sick relatives in case sanctions ended one day and they could find a cure. Not far from Buha'a Hussein's fruit groves, Ali Ahmed Suwaidan, who lived beside a canal, had just had such an X ray. It showed the head of his five-year-old daughter Fatima, who was playing at his feet. He explained: "There is something wrong with her balance. She cannot stand up." He held her upright for a moment and then removed his hands. Fatima immediately crumpled to the ground at his feet.

On the banks of an irrigation canal not far away, a lean-looking woman in dark peasant clothes named Nahay Mohammed was clambering down to get water with a steel bucket attached to a piece of rope. "It is bad water, of course," she said. "It gives you stomach pains and hurts the kidneys, but the purified water supply was cut off in 1991." Heliathan Alwan, a farmer from the same village, said he had recently visited the nearest town to see if they could restore the drinking water, but was told it was impossible.

Whatever sums might still be kept in reserve by Saddam for his private and military purposes, it was clear that the resources required to feed the country were running out. In 1994, food rations were cut back. In 1996, the merchants of Amman noticed that the gold ingots brought for sale by Iraqi government officials had changed. Previously they had been of the standard shape and composition of ingots made for central banks. Now, close examination indicated that the gold bars had been made from melted-down wedding rings and other jewelry.

The blockade on the economy was unrelenting. Every single item legally exported to Iraq had first to be submitted for approval to the sanctions committee operating under the auspices of the Security Council. This committee was rigorous in excluding even the most inoffensive items suspected of being of "dual use," with military applications. Apart from the chlorine excluded because of its possible use in manufacturing chemical weapons, items denied included spare parts for ambulances—because they could be useful for transporting troops—and lead pencils—the graphite could have a nuclear application. Bedsheets were denied in one case, as were exercise books. Tires, which could certainly be put to use by the military, were absolutely embargoed, and Professor Jawad's Subaru spent more and more time off the road. Even when indulgent, sanctions approval was a slow-moving process. Often it took a year for permission to import a spare part.

The suffering caused by sanctions did produce some official action on the part of those enforcing them. In the summer of 1991, the Security Council offered an "oil-for-food" deal. As originally formulated, Iraq would be allowed to export $1.6 billion worth of oil every six months. The money would be paid into an account controlled by the United Nations and spent under UN auspices on food and medicine after approval of such items by the sanctions committee. It appeared, and may have been designed, to show generosity and humanitarian concern on the part of the victors, though why the amount of money offered was limited, given that the United Nations would control the spending of it, was never explained.

Even so, $1.6 billion would have made a difference. Saddam, however, obliged his enemies by rejecting the offer on the grounds that its strict requirements for UN supervision of oil revenues infringed on Iraq's sovereignty. He continued to reject it on the same grounds for the next four years. In 1995, the Security Council, recognizing that the food situation in Iraq was growing worse, adopted Resolution 986, an improvement on the earlier offer in that Iraq was now allowed to earn $1 billion every ninety days, though much of that money was to be diverted as compensation to Kuwait and to pay the United Nation's own bills. Citing the sovereignty argument, Iraq continued to object before finally accepting the deal in May 1996, not long after the World Health Organization reported that "The vast majority of the country's population has been on a semi-starvation diet for years."

Oil began to flow at the end of the year, and in March 1997, the first shipment under the oil-for-food agreement, a load of chick peas and white flour from Turkey, arrived in Iraq. Food became more plentiful, but by the end of the 1990s, after eight years of sanctions, the Iraqi economy could not be restored by humanitarian aid. "The infrastructure is collapsing and it will take ten to twenty years to restore," said Denis Halliday, a fifty-seven-year-old Irish Quaker. Appointed UN humanitarian coordinator for Iraq in August 1997, he was responsible for monitoring expenditures now available under the oil-for-food arrangement. He cited as one example the fact that, even with the money now available, much of the electric power system was beyond repair. "We have generators that are twenty years old. When we go to the manufacturers [we find] they don't make the spare parts anymore." His office estimated that $10 billion was needed to restore Iraq's electrical system, but only $300 million was available even under the terms of an expanded oil-for-food program agreed upon between Iraq and the UN in early 1998.

By the end of the 1990s, the Iraqi economy was breaking down everywhere. Just as the UN Security Council was announcing in February 1998 that Iraq could in future export $5.2 billion worth of oil every six months, Hussein Ali Majhoul, an eight-month-old baby, was dying in the al-Khatin Hospital from infectious diseases contracted in the southern outskirts of Baghdad. Beside his bed was an empty oxygen bottle. "He has meningitis," said Dr. Deraid Obousy, the weary-looking director of the hospital, gently pressing the side of Hussein's neck. "He is already unconscious. It is in the hands of God. We don't have any more oxygen bottles in the hospital and we don't have any more money to hire a truck to pick up a new one from the factory that refills them on the other side of Baghdad." In the forecourt of the hospital, there were what from the distance looked like a fleet of trucks. But on closer inspection all turned out to be without wheels, their axles resting on stones, or missing essential engine parts. They had been progressively cannibalized over the previous eight years to try and keep one vehicle on the road.

Dr. Obousy, forty-six but looking older, was gloomily reading an old copy of the *British Medical Journal* that had found its way to Baghdad despite sanctions. He said that in Britain, where he had worked in hospitals for four years, "a place like this would definitely be closed. They would say it is rubbish. It is getting into the hot season and we have no mosquito netting for the windows, or air conditioners, or even enough sheets for the beds." All this was confirmed by a tour of the wards. The smell of disinfectant did not quite mask the stench of the lavatories. The patients were eating a meager meal of rice and chick-pea soup. Sitting beside baby Hussein Ali's bedside were his mother, Nada, and her husband, Ali, a factory worker, who explained that their family income was about 14,000 Iraqi dinars ($10) a month, which had to support both them and their parents. Nada was pretty and slim. "She is obviously malnourished herself," explained the doctor. Later in his office he spoke of the poverty of his patients. He admitted that he did not officially earn more than $10 a month himself. He had to have a private clinic to keep going, but this was producing less income because his patients had progressively sold their possessions and now had no money with which to pay him. His own TV and radio had long since been sold to buy food.

In the Canal Hotel, the UN aid headquarters, Denis Halliday was appalled by what he found in Iraq. His career had been spent in the UN Development Program, trying to build up the resources of impoverished countries. Now he was in charge of funneling aid to a country being ruined by UN sanctions. "You go to schools where there are no desks," he said. "Kids sit on the floor in rooms that are very hot in summer and freezing in winter." Overall, Halliday thought that humanitarian aid was "only Band-Aid stuff"—a point reinforced by the fact that even after the implementation of the oil-for-food agreement, levels of malnutrition remained unchanged. He said the only real solution was to "lift sanctions and pump in money." For this humane official, it came down to a moral argument. Early in his tenure, Halliday had remarked that sanctions were "undermining the moral credibility of the UN" and "in contradiction of the human rights provisions in the UN's own charter."

Back in July 1991, when the notion that the blockade of Iraq might go on for years and years seemed incredible to anyone aware of conditions in the country, the relief worker Doug Broderick had made the shocking observation that as many as 175,000 Iraqi children could die because of the public health conditions. He called it a "disaster in slow motion." Seven years later, his prediction had been proved wrong. Not 175,000 children had died, but upward of half a million. By the end of 1995 alone, according to an investigation by the United Nations Food and Agriculture Organization, as many as 576,000 Iraqi children had died as a result of sanctions. The World Health Organization, citing figures from Iraq's Ministry of Health, estimated that 90,000 Iraqis were dying every year in Iraq's public hospitals above and beyond the number who would have expired in a "normal" situation. The precise number was not exactly known because many Iraqis had stopped using the health care system.

Broderick had, however, been correct in calling the situation a slow-motion disaster. At the end of the Gulf War, the Western public had been moved to pity by reports of slaughter on the "road of death" leading north from Kuwait City. An Iraqi convoy of hundreds of trucks and cars fleeing up the highway to the border had fallen, easy prey to allied warplanes; disquiet evoked by the "turkey shoot" had helped impel Bush to order a cease-fire. In fact, the casualties on the road had been comparatively light—perhaps four or five hundred—as was the entire Iraqi death toll from the fighting and bombing. The real slaughter came later, but because it happened in slow motion, without arresting images of victims with protruding rib cages or heaps of corpses, the impact in the West was minimal. Dry statistics detailing remorselessly escalating infant mortality rates, or the percentage of underweight children, or even the death of little Hussein Ali Majhoul for want of a working truck to drive across town could not jump-start an international furor over the sanctions policy.

For every cited statistic on infanticide, the enforcers of sanctions could point to the latest evidence of Saddam's perfidy in concealing his meager stockpile of deadly weapons. In 1996, CBS News' *60 Minutes* broadcast a chilling exchange. Correspondent Lesley Stahl interviewed Madeleine Albright, then U.S. ambassador to the United Nations. Albright maintained that the sanctions had proved their worth because Saddam had made more admissions about his weapons programs and because he had recognized the independence of Kuwait (which he did in 1991, right after the war).

"We have heard that half a million children have died. I mean, that's more children than died in Hiroshima," said Stahl. "And, you know, is the price worth it?"

"I think this is a very hard choice," replied Albright, "but the price—we think the price is worth it."

Insofar as there was a debate, each side resorted to tendentious arguments. Those moved by the plight of ordinary Iraqis derided the often fruitless efforts of Unscom inspectors to find Saddam's remaining cache of weapons. Unscom officials, on the other hand, sincerely believed that the degree of suffering in Iraq was being deliberately exaggerated by the government and that those who raised the issue were dupes. As one inspector, a veteran of many inspection missions to Iraq, remarked to us, "Those people who report all those dying babies are very carefully steered to certain hospitals by the government." It was impossible to convince him that hospitals like the one in which Dr. Obousy worked were not in short supply.

For Washington, sanctions, as a former senior CIA official observed in early 1998, had been a "demonstrable success." By this he meant that they had kept Saddam too weak to reassert himself as a power in the region. By this yardstick, the policy had indeed justified itself. The leader of a country where over a quarter of the children were "stunted," where the once flourishing and highly educated middle class had been utterly ruined, that overall had sunk from the economic and social level of Greece to that of the barren sub-Saharan wasteland of Mali was now obviously less of a threat.

If, however, the goal of the sanctions policy had been to actually bring down Saddam, it had failed demonstrably. Indeed, in some respects, sanctions had actually strengthened the dictator's position. The agony of ordinary Iraqis, fitfully reported in the media, may have had little resonance in the United States, but in the Arab world it was a different matter. Even those most disposed to fear and loathe Saddam were moved by the plight of their brethren. Prince Khalid bin Sultan, nephew of the king of Saudi Arabia and commander of the Arab forces in the Gulf War, called for an end to sanctions on Iraq because "they have only reinforced President Saddam Hussein's hold on power while starving [the] Iraqi people." It was a sentiment that was to cost the United States dearly, as we shall see.

Iraqis themselves implicitly confirmed the prince's point by unequivocally blaming the United States, rather than Saddam, for their troubles. Ali Jenabi, a highly educated economist and friend of Professor Jawad, exclaimed angrily: "Do you think that Britain and the U.S. are really afraid of our biological weapons? Of course they are not. The sort of things we have, any country could make in a bathtub. A single religious maniac in Japan was able to make nerve gas. They just want to keep Iraq weak and divide up its oil."

Rhetoric aside, Saddam Hussein himself showed little sign that he was moved by his people's plight. To him, as to those who were enforcing the sanctions, they were hostages, bargaining chips whose very suffering was an asset. Thus he would arrange displays of dead children to shame his enemies, as he did with Rolf Ekeus, into giving him free rein in his drive to rebuild his power, perhaps with the remnants of the weapons programs he had managed to keep hidden. But the dead children were real. The tragedy was that in aiming at the hostage taker, the United States and its remaining allies were killing the hostages.

Meanwhile, all observers agreed, Saddam Hussein and his family were immune to the deepening privation outside the gates of the new palaces that the leader set to building with whatever scarce building materials were available. Secure in its personal comforts, this family was free to pursue its own dark and bloody intrigues.

SIX

Uday and
the Royal Family

It was a cold night in mid-February 1992. Across Baghdad and the rest of Iraq, Saddam Hussein's subjects were sinking ever deeper into miserable privation. The eerie yellow fog that had mingled with smoke from the burning oil refinery during the war a year before was again swirling across the city, over the pools of sewage lapping at doorways in the poverty-stricken southern suburb of Saddam City, over the gardens of once-affluent middle-class villas in al-Mansour, where householders contemplated which family heirloom they would next take to the market, over the forecourt of the high-rise al-Rashid Hotel and the fleet of expensive sport utility vehicles that had just pulled up at the entrance to the lobby.

The fog and the misery stopped at the doors of the hotel. Inside, the National Restaurant, one of the most expensive establishments in Baghdad, was having a typically busy evening. Seated amid the ornate decor of casbah gold leaf and black lacquer, the new rich and their families were eating their fill of the lamb kebabs, thick steaks, and grilled Tigris fish rushed by busy waiters onto the crisp white cloths of the crowded tables. These customers were, for the most part, the smugglers and profiteers for whom the shortages caused by sanctions had provided boundless opportunity, now carelessly lavishing a month's income or more for an ordinary family on their evening's entertainment. Amid the rattle of plates and roar of conversation came the sound of Mr. Abdullah, the house musician, plucking the strings of his santir.

Suddenly, there was a commotion at the door. A new party had arrived, the occupants of the motorcade that had just drawn up outside. Some of them were dressed in the black leather jackets that security agents of the regime had adopted as their trademark. These fanned out to various corners of the spacious dining room, eying all before them with the attentively suspicious gaze of bodyguards. Two of them disappeared through the doors into the kitchen. Close at their heels strolled a group of casually well-dressed men, most of whom hung back with a suggestion of deference to the two youngest among them. These two were both dark, one with a distinctive designer stubble, the other bearing a softened version of the face that adorned public posters the length and breadth of Iraq.

The new arrivals brought a perceptible shift in the level of background noise in the room. No one was foolhardy enough to stare directly at the party that was now heading for a corner table next to Mr. Abdullah, but everyone was conscious of their presence. Finally someone hissed, "It is the sons of the president!" The young man with the stubble was Uday, at twenty-eight the eldest son of Saddam Hussein. He and his brother, Qusay, were seating themselves at the head of the table, chatting relaxedly with the rest of the company.

With the possible exception of his father, there was at that time no individual in Iraq more universally loathed and feared than Uday. Everyone knew the stories of his greed, extreme brutality, and taste for public violence. A few days earlier, Mohammed Jawad had beckoned two guests into his garden to whisper that "just down the road," Uday maintained his own currency printing plant "working around the clock." Few Iraqis had not heard of the occasion, six years before, when an army officer had attempted to defend his girlfriend from Uday's advances in a discotheque. Uday had shot the man dead on the spot. Taxi drivers feared even to drive on the street in front of his office.

To the untutored eye, it might have seemed that Uday and company were merely a convivial and inoffensive group of friends out for an evening meal. But within minutes of their arrival, the busy room had been entirely transformed. Now a sea of deserted tables stretched around them, some still littered with half-eaten dishes. The restaurant's Iraqi patrons had abandoned their dinners and hurried away, anxious to escape the lethally dangerous individual who had appeared in their midst.

Mr. Abdullah struck up a new tune. Soon the party in the corner was joining in a rousing chorus from, as their host later explained, "The days of the Thousand and One Nights," Uday beating time with a large Havana cigar. Johnny, the suave Sudanese maître d', bustled up and down the table lighting cigarettes and refilling glasses from the bottles of Black Label and champagne that lined the cloth. Uday himself drank only from a decanter of cognac he had brought with him. The atmosphere gave little hint of menace, save for the beads of sweat on the waiters' brows. By and by, one of the party stood up and made his way over to where two journalists from America (Leslie and Andrew Cockburn) were sitting, transfixed by the scene before them. He was carrying a bottle of champagne. After filling their glasses and introducing himself as "Ahmed," he waved at the table behind him. "There is a lion in Iraq," announced the champagne bearer in slightly slurred tones, "and these are his cubs."

Saddam Hussein has always loved his cubs. The only joke the Iraqi leader has ever been heard to tell is an affectionate quip to the effect that Uday has been a "political activist since he was a baby." As he tells it, his wife, Sajida, would visit him when he was in prison in Baghdad in 1964, bearing baby Uday, who had been born on June 18 that year. They had married the year before when Saddam returned in triumph from exile in Egypt as soon as the Baath Party overthrew and killed President Qassim. Sajida was the daughter of his uncle Khairallah Tulfah, who brought him up, and their marriage had been arranged when he was five years old. A rare photograph of the newly married couple, in which Saddam is for once clean-shaven and without his luxuriant mustache, reveals that the first cousins even look alike. Each has the same slightly pursed lips and large, deep-set brown eyes that stare coldly at the camera.

Sajida was visiting her new husband in jail because a year after the wedding, the Baath Party was out of power again and Saddam, one of its rising leaders, had been arrested. The punch line of Saddam's story, as he told it in later years, was that Sajida would hand him the baby (the guards saw nothing suspicious in this touching sign of paternal affection). As Saddam cuddled his firstborn, he would slip a hand under the diapers, where comrades in the Baath Party, still free, had hidden secret messages for their imprisoned colleague.

As Uday grew through childhood, the affectionate bond evidently persisted. Family snapshots show father and son playing together on the beach. Nor was this affection confined to Uday. Qusay, born in 1966, also appears in the beach games, and Saddam always presented himself as a besotted and overtolerant father to his three daughters, Raghad, Rina, and Hala. In his first and only interview about his family, with the Iraqi women's magazine *Al-Mar'a* in 1978, Saddam said: "When they were children, I loved my daughters most, beginning with Raghad." When she was a pretty ten-year-old with light brown hair, he had a picture of himself taken mending the sleeve of his eldest daughter's flowery pink dress with a needle and thread.

By the time he gave that interview, Saddam was on the brink of supreme power, which he achieved the following year with the elimination of his rivals in the leadership of the Baath Party. Such unchecked power had not been seen in Baghdad since the days of the medieval caliphs, and, as in the medieval kingdom, the ruler's extended family constituted a court, with princes, great nobles, and lesser lords. Uday himself was often referred to as "The Prince" by Iraqis.

The higher nobility came from two branches of Saddam's extended family, or clan, the Bejat, which in turn formed part of the Albu Nasir tribe. The first of these branches were his al-Majid cousins, nephews of his father, Hussein al-Majid, whom he never knew. They played a public and aggressive role in supervising the army and repressing the Kurds and the Shia. The second branch of the family on which Saddam relied were the Ibrahims—his three half-brothers, Barzan, Watban, and Sabawi, sons of his mother, Subha, by her second marriage to Ibrahim al-Hassan. They played a critical role in the intelligence and security services. Though temporarily eclipsed following Subha's death in 1983, they again returned to prominence after the Gulf War.

For Saddam, the bonds forged by the blood ties within his extended family were not close enough. To create even tighter links, he resolved that the cousins be further united through matrimony. Marriage between first cousins is common in Iraqi tribal society, and the beloved daughters Raghad and Rina were accordingly bestowed on their al-Majid cousins, Hussein Kamel and Saddam Kamel, both rising stars within the family in the 1980s. Barzan married Ilhan, the exceptionally good-looking daughter of Khairallah Tulfah, Saddam's uncle. Uday married Barzan's daughter Saja after a brief marriage to the daughter of Izzat Ibrahim al-Dhouri, vice chairman of the Revolutionary Command Council and leader of the powerful Dhouri clan, long allied to Saddam.

As rebellion exploded in the south of Iraq and Kurdistan in early 1991, Saddam looked first to two of the al-Majids, Ali Hassan al-Majid and his nephew Hussein Kamel. On March 5, he put Ali Hassan, fifty years old and until recently governor of Kuwait, in charge of security as interior minister. A diabetic with a menacing, rodent-like face and a scraggly mustache who suffered from hypertension and spinal infections, he was, above all, the family enforcer, though this title had other, well-qualified contenders.

It was a reputation he had won in Kurdistan. A former army driver and NCO, he had presided over the regime's greatest crime. In 1987, he was appointed secretary general of the Baath Northern Bureau in charge of suppressing the Kurds, who had taken advantage of the Iran-Iraq war to rise in rebellion. Over the next two years, he slaughtered, using poison gas and execution squads, between 60,000 and 200,000 Kurds. Much of the region became entirely depopulated. When the Kurds rebelled again in 1991, they captured Iraqi security archives, including tapes of Ali Hassan addressing subordinates in his distinctive, high-pitched, whiny voice as he exhorts them to further atrocities. At one moment he is heard rhetorically responding to potential critics of his execution of Kurdish men, women, and children in 1988. "Am I supposed to keep them in good shape?" he asks rhetorically. "No, I shall bury them with bulldozers."

In another tape he is heard telling Baath Party cadres to disregard any international reaction to the use of chemical weapons against the Kurds: "Who is going to say anything? The international community? Fuck them."

Soon after his appointment as interior minister in 1991, Ali Hassan was part of an Iraqi delegation that met with Kurdish leaders. He appeared nervous. The Kurds said that by their calculations, at least 182,000 Kurds had disappeared during his two years in charge of Kurdistan. Ali Hassan sprang to his feet and angrily shouted: "What is this exaggerated figure of a hundred and eighty-two thousand? It couldn't have been more than a hundred thousand." A video of him interrogating and beating Shia prisoners captured in the south in March 1991(see Chapter One) showed that he had not changed his methods.

The second of the al-Majid cousins to whom Saddam turned to defend his power during the rebellions was Hussein Kamel, the thirty-seven-year-old husband of his favorite daughter, Raghad. Kamel's earlier services to the regime had, if anything, been even more significant than those of his uncle Ali Hassan.

In the terrible month of March 1991, with the regime tottering on the brink of destruction, Hussein Kamel had been in the thick of the fighting, leading the Iraqi armored columns in their assault on the holy city of Kerbala. At the end of the battle, he had marched into the badly shattered shrine-tomb of Imam Hussein, the seventh-century Shia martyr and saint, shouting triumphantly: "We are both called Hussein and I have won." Rumor quickly circulated news of Hussein Kamel's defiance of the founder of Shiism in Iraq. The fact that he did not even remove his army boots when entering the shrine was cited as an example of his arrogance and contempt for the Islamic tradition to which 55 percent of Iraqis belong. Kamel himself later had doubts about the wisdom of his actions. When he was diagnosed as having a brain tumor in 1994, he believed it was because he had profaned the shrine. Returning from a successful operation in Amman, Jordan, he diverted his ambulance so he could pray at the tomb of Imam Hussein to give thanks for sparing his life.

Imam Hussein was not the only target of Hussein Kamel's ruthless arrogance. Professional army officers resented him because of his rapid promotion and inexperience. In 1982, he was only a captain in the army but was given the job of forming the elite Republican Guard divisions to spearhead the counterattack against Iran. He was promoted to lieutenant general and put in charge of military procurement in 1988. He showed great energy in both jobs, as well as extreme greed in seeking commissions on military contracts. When a scheme to build an oil pipeline to the Red Sea, which Hussein Kamel supported, was defeated by a rival scheme, backed by Barzan, Saddam's brother, in Geneva, he launched an anticorruption drive. The deputy oil minister and Nazir Auchi, a prominent Iraqi businessman, both linked to the successful bidders, were accused of paying bribes and summarily executed.

Like almost all of Saddam's close relatives, Kamel first made his mark running one of the security services. In his case, he helped organize the Amn al-Khass, a special inner-security agency around the president, founded after an attempt to assassinate Saddam in the mid-eighties. But he believed his chief claim to fame was to create the elite Republican Guards out of a single brigade that had suffered heavy casualties fighting the Iranians. "One of the regiments had just six soldiers," he would recall. "The second regiment consisted of only twenty-four soldiers." It was his proudest moment. He was given the right to co-opt any officer he wanted out of the regular army. Within a few years, he had built up the Republican Guard until it had thirty-seven brigades and was the main strike force of the Iraqi army.

In sharp contrast to Uday, Kamel was a puritan. He drank no alcohol or even tea, an astonishing exercise of self-denial in Iraq, where social and business life is punctuated by regular consumption of little glasses of sweet tea. He was often aggressive or petulant but gave the impression of being brittle under pressure. Strident in his attacks on others, he was wounded by criticism of himself. In 1991, his remarks at the regional Baath Party congress so angered other leaders that five of them walked out of the meeting. A few days later, he resigned as defense minister and, when asked to reconsider, not only refused, but, as he later admitted, "did not go to the office for three months."

Hussein Kamel's authority, like that of others in the family circle, depended on access to Saddam. Barzan, the Iraqi representative to the United Nations in Geneva from 1988 to 1998, later claimed that Kamel "threw a ring around the president and prevented others from getting to him." He went on to say that Saddam "relied on him in a major way even though he was not competent as a military man, engineer, or politician. In 1975, he was a driver in the president's motorcade, and later the president gave him promotions he did not deserve. He got into a position where he could see [the president] night or day."

Kamel's rise had provoked the jealous rage of the Ibrahim branch of the family since the early 1980s. Barzan, intelligent and articulate, in appearance a slimmed-down version of Saddam, played a critical role as security chief in the Iraqi leader's rise to power. Both he and his two brothers fiercely resisted the marriage between Kamel and Raghad in 1983, when she was barely sixteen. They correctly foresaw that this would dilute their own power. A dozen years later, Barzan was still fuming that Hussein Kamel's "only legitimacy is that he married the president's daughter. Otherwise nobody would have cared about him. He now speaks about his clan, but within that clan there is an entire generation that supersedes him. He is a rash, aggressive, and hard boy who knows no courtesy."

To add insult to injury, Saddam Kamel, Hussein Kamel's younger brother, later married Rina, Saddam Hussein's second daughter. Saddam Kamel, an officer in the security service, was always overshadowed by his sibling. His chief claim to fame among Iraqis was his starring role in *The Long Days*, the epic depicting Saddam Hussein's assassination attempt on President Qassim.

Barzan's description of Hussein Kamel's lack of qualifications was accurate, but in the face of the uprisings, Saddam probably felt he needed the straightforward energy and brutality of his Majid cousins. Meanwhile, however, his Ibrahim half-brothers had recovered some of the influence they had lost in the eighties. They were assigned to low-profile but crucial posts dealing with intelligence and security. Control of this field had always been Saddam's preeminent tool in building and maintaining his power, and in the postwar years he moved to tighten the family hold over these vital organs. Sabawi, Saddam's youngest half-brother, was made head of General

Security (al-Amn al-Amm). Qusay, of course, in his overall supervisory role of security overseer, had an even more powerful position.

In November 1991, Hussein Kamel left the Defense Ministry, which he had headed since the immediate aftermath of the war, and returned to military procurement. There was no sign of his losing influence. His defense post was taken by Ali Hassan al-Majid, whose place at the Interior Ministry was in turn filled by the ruler's half-brother, Watban Ibrahim.

This game of musical chairs among Saddam's inner circle may have been a recognition that the nature of the threat to the regime was changing. By the late summer of 1991, the army, after heavy losses, had withdrawn from Kurdistan. It dug in along a fortified military line, which snaked across the plain below the Kurdish mountains like an Iraqi version of the Maginot Line. Ground fighting had ceased everywhere in Iraq, apart from occasional clashes with Shia guerrillas in the reed beds of the southern marshes.

The threat of armed insurrection died away, but Iraqis realized that the siege of Iraq instituted by the international community after the invasion of Kuwait was not ending. Sanctions remained in place. There were almost no oil exports. Political isolation was actually increasing. The Soviet Union, Iraq's old ally, collapsed in 1991. Even Jordan, a friendly neutral during the Gulf War and Iraq's one legal means of access to the outside world (as opposed to the smuggling routes across Kurdish-held territory), was beginning to reassess its links with Baghdad.

At home, there was the serious threat that the powerful Sunni tribes, a key component of Saddam's power base since the earliest days, might also be reassessing their links with the regime. In the days when Saddam ruled an Iraq that was united, fabulously wealthy from oil revenues, and a growing regional military power, they had been happy to support the man to whom they were, in any case, linked by tribal alliances. They had rallied to Saddam in the face of the Shia and Kurdish rebellions that threatened Sunni control of the country.

Once the rebels had been defeated, however, the costs to these tribes of continuing rule by Saddam began to outweigh the benefits. The United States had made it clear that sanctions would continue as long as Saddam Hussein remained in power. Sanctions were rapidly impoverishing the mass of the population, including not only the Shia, who were almost entirely excluded from the ruling apparatus, but also Sunni technocrats, Baath Party officials, and army officers. It would have been surprising if there had been no reaction to this situation and, indeed, in the next few years there were a series of conspiracies against the regime among officers from Sunni Muslim tribes, who had traditionally supported it. The Juburis, for example, centered around the northern city of Mosul, had prospered under Saddam's rule, with many of their members reaching high positions in the military and security services. Nevertheless, in 1993, two senior Juburi air force officers—the deputy commander and the director of operations—were arrested for attempting a coup. There was further unrest among senior officials from other important tribes such as the al-Dhouris and the Dalaim. The danger for Saddam was that the plotters came from cities like Ramadi, Mosul, and Samarra, which had supported him in 1991. In a 1992 speech, Saddam mocked "imperialists" for recruiting to their conspiracies "treacherous and perfidious people who have spent part of their life in Tikrit." He could no longer take for granted the loyalty of the lesser lords who had once clustered around his throne. Those in Washington who believed that there was a good chance that the Iraqi military would oust Saddam if the United States could keep up the pressure were not being wholly unrealistic. After all, one significant reason for government control having evaporated so quickly during the northern uprising was the sudden defection of formerly loyal Kurdish tribes, well armed and organized into the Jash militia by Saddam, to the rebels.

The Iraqi leader had to prevent the Kurdish example happening again elsewhere. While the regime had always drawn its strength from the Sunni towns on the upper Tigris and Euphrates, where tribal links were strong, now Saddam did everything to conciliate tribal leaders around the country. Cars were given to Shia tribal sheikhs in the south, who were also invited on trips to Baghdad with all expenses paid. In 1992, these provincial dignitaries could be seen uneasily sampling the modern conveniences of the al-Rashid Hotel, ascending and descending in the elevators as they strove to master the controls. Irreverent foreign journalists dubbed them "the flying sheikhs." Prohibitions against the concentration of land holdings, introduced after the 1958 revolution, were dropped. In 1992, Saddam even apologized to tribal chiefs in southern Iraq for past agrarian reforms. The government paid high prices for agricultural products, making wealthy the many tribal leaders who owned land. People in Baghdad began to notice that those who now filled the city's restaurants were increasingly tribal notables, wearing traditional flowing dishdashes and parking their shiny new four-wheel-drive vehicles outside.

The heightened profile of these traditional groupings was paralleled by the declining importance of the Baath Party organization, which between the revolution of 1968 and the Gulf War had tightened its grip over Iraqi civil society by controlling all civic organizations. Faleh Jabber, formerly a leading member of the once powerful Iraqi Communist Party and an acute observer of Iraqi politics, cited a telling example of these two trends in a perceptive essay. "The telegrams of support sent to the President on Army Day and National Day are no longer from trade unions, students' organizations, professional societies, political parties or other modern social groups," he wrote in 1994. "Nowadays they are signed by tribal sheikhs whose tribes are named and even the number of their tribe members is given. The revival of old social classes seems clearly intended to forge new social alliances, particularly in the south."

Saddam might attempt to conciliate and reward his grassroots supporters but, as indicated by the periodic conspiracies (all of which proved abortive), he could never depend on them. Therefore, to defend himself and his regime, he turned more and more to his immediate family. Cousins such as the fearsome Ali Hassan al-Majid and Hussein Kamel had already proved their worth, but, as the base of the regime appeared increasingly insecure, Saddam reached for the one man in Iraq whose reputation for personal violence and cruelty was as great as his own: Uday, the young prince.

Everything about Uday was flamboyant. Other members of the ruling family lived in the shadows, but Uday was to be seen in Baghdad's best hotels, restaurants, and nightclubs. In the early 1990s, his headquarters was a ten-story yellow building in East Baghdad, with medieval-style watchtowers for machine gunners on its outer walls, that housed the Iraqi Olympic Committee, of which Uday was chairman. It was probably the only Olympic headquarters in the world to have its own prison.

Uday was a physically striking figure. His enormous staring brown eyes dominated his face and he usually had a five days' growth of beard. In a photograph taken in 1977, when he was thirteen, he wears a loud striped jacket and an enormous black bow tie. The impression given is of somebody trying to assert his personality against almost overwhelming odds. His schoolmates speak of him as rarely turning up for classes, and even then five bodyguards accompanied him to the classroom. Nevertheless, he and his brother, Qusay, also said to have flunked in school, both learned fluent English. Uday even had an early ambition to be a nuclear scientist. He spoke about this when he was sixteen and repeated the story of his thwarted hopes in a conversation with Leslie Cockburn in Baghdad in 1992. He said he had traveled to the United States in search of further education: "I did my SATs, everything. I did very well. Passed with high marks." But, he claimed, his ambitions were thwarted by the outbreak of the Iran-Iraq war. He said: "You see, I wanted to do nuclear studies, and at that time there was a problem with Iraqis doing that." It was a bitter blow. "I wanted to go to MIT," he recalled sadly.

Some aspects of his early education were, at least by his own account, unique. In the late 1970s, Saddam sent Sajida and the boys off for a family vacation in Spain. They stayed with General Hassan al-Naquib, who later defected to the opposition but was then the Iraqi ambassador to Madrid. Al-Naquib had two sons of roughly the same age as the Saddam youngsters and also a young daughter. As the four boys played together, Uday boasted that his father would send them to the prisons to witness torture sessions, in order to prepare them for "the difficult tasks ahead." In an apparent effort to impress the little girl, he threw in the further detail that they were sometimes allowed to execute prisoners themselves. Given the general ambience of the Saddam family, this may even have been true, and, in any case, Uday's choice of themes for playground braggadocio are in themselves illuminating. There is certainly evidence that in 1979 the boys were treated to an outing to the semipublic executions of Baathist leaders opposed to their father. Whether because of the childhood experiences or not, Uday always relished instilling fear as well as displaying a taste for very public violence, especially when drunk on whisky or cognac. An Iraqi who used to work for him remembers going with Uday to a nightclub. As he described the evening, "Uday lined up a group of male gypsy singers onstage, told them to drop their trousers and sing while he fired his submachine gun just over their heads. After ten minutes, some began to urinate with fear. Then he told them all to get dressed, gave them some money, and told them to go home."

At the Saddam University of Technology in Baghdad, founded by the Baathists, Uday led the cosseted life of a crown prince. He was usually called "Ustav Uday," meaning Master Uday. He joined the Baath Party at the age of twelve. Mohammed Dubdub, the chairman of the National Union of Iraqi Students (known as bear-bear because his name as written in Arabic sounds like the word for "bear"), became his political tutor. When the Iran-Iraq war began in 1980, Uday used to go to the front. He was always accompanied by the Iraqi chief of staff, General Abdul Jabber Shanshal, who walked respectfully behind him. While his father was careful not to put Uday's life in danger, he wanted to show that his son was doing his bit, so Uday went through flight training to learn to fly an army helicopter.

In 1982, father and son, accompanied as usual by the dutiful chief of staff, turned up near Basra. Wafiq al-Samarrai, then a rising star in military intelligence, happened to be on the scene and witnessed the performance put on by father and son for the edification of the troops. There was heavy fighting just to the east and Saddam loudly ordered his firstborn to go and attack the enemy. On cue, General Shanshal begged Saddam not to send Uday on this dangerous mission. "But Uday got into his helicopter and we could see him firing his missiles. Later it turned out that he had hit our troops," recalled al-Samarrai with a laugh. "One person was injured. The unit involved even sent a report saying: 'You should punish the pilot.'" Only years later did al-Samarrai learn from Hussein Kamel that Saddam and Uday had choreographed the whole event, including the firing of missiles at a safe distance from the Iranians, before they had left Baghdad.

In the final years of the war, while cousins like Hussein Kamel were performing important functions such as superintending the Iraqi nuclear weapons project, Saddam began permitting his eldest son to play a minor political role. Uday took over the Iraqi Olympic Committee in 1987 and turned it into the Ministry of Youth (there was another, real minister of youth who received a medal from Saddam for proposing the abolition of his own ministry "in accordance with Baathi values"). In the coming years, Uday was to use the Olympic committee as a base for involvement in all aspects of Iraqi life. He seemed to have unlimited funds, employing any veteran officer who applied for a job. He set up and managed football teams. Players who failed to score goals or prevent opposing sides from doing so were routinely sent to jail. Spectators could tell who had been punished because the player-prisoners had their heads shaved by their jailers.

Uday's reputation as a brutal playboy was known to the Iraqi elite, but even a society as used to violence as Iraq was shocked and astonished to learn in November 1988 that he was in prison for the public murder, during a party on an island in the Tigris, of one of his father's closest aides.

The motive for the killing was as lurid as the murder itself. Saddam had married Sajida Tulfah in 1963, but had mistresses, such as Majida, the wife of Hamed Youssef Hamadi, the minister of information. But in the middle of the Iran-Iraq war, Saddam fell in love with Samira al-Shahbander, a beautiful ophthalmologist. They were secretly married and produced a son, Ali. Sajida was enraged. She was not only his wife, but related to important members of his clan. Her brother, Adnan Khairallah Tulfah, was Saddam's boyhood friend and minister of defense. In her fury, Sajida sought out Uday, her favorite child. The cause of her grief, the man responsible for arranging his father's illicit liaison, she said, was none other than Kamel Hannah Jajo, who for years had been practically a member of the family as Saddam's aide, bodyguard, and sometime food taster.

On October 18, unaware of or oblivious to the seething passions he had aroused, Jajo threw a party on an island in the Tigris called Umm al-Khanazir (Mother of Pigs), not far from the presidential palace on the west bank of the river. It was a grand occasion, with Suzanne Mubarak, the wife of the Egyptian president, as guest of honor. Uday was not invited. Considering this a slight, he decided to hold his own rival party next door. Only a low hedge separated the two venues. The exact details of what happened next come from Latif Yahia, who, because he closely resembled Uday in appearance, had been recruited as his double the year before to stand in at official events. As such he became, at least for a while, a member of Uday's social circle, a dubious privilege.

According to Uday's double's account of the dramatic events of the evening, Uday was looking for a confrontation but did not want to make the first move. He told Adel Akle, his favorite singer, not to play too loudly and to merely provide background music. As the evening wore on, Uday became very drunk, mixing straight whisky and cognac at the buffet. At about midnight, shots rang out from the other side of the hedge. This was Jajo, in a typical Iraqi gesture, firing volleys into the air with his submachine gun. He too had been drinking heavily. Uday sent an aide to tell him to stop making so much noise.

When the aide returned, he told Uday that Jajo not only refused to do anything about the noise but had sent back a message: "Kamel Hannah says he obeys only the president's orders." Enraged, Uday forced his way through the hedge. All evening he had been carrying a battery-powered electric knife, called a Magic Wand, which he normally used for cutting his roses. As he drank, he had been nervously switching it on and off, slicing up fruit, napkins, and, at one stage, even his cigars. When he appeared at the second party, Jajo was standing on a table, still holding his gun in one hand and a spare clip of ammunition in the other. Uday shouted at him: "Get down." Jajo did so, but then repeated: "I obey only the commands of the president."

Uday lashed out and hit him twice on the head with the electric knife and, as he staggered back, hacked at his throat. Jajo lay on the ground, trying to pick up the submachine gun he had dropped. Uday kicked it to one side and shot him twice with his pistol. He then walked back through the hedge to his own party, but left immediately to lock himself in a room in a nearby government building. Some of the officers at the party rang the palace and Saddam Hussein himself arrived within a few minutes, wearing just trousers and a shirt, his feet thrust into shoes without any socks. When an ambulance arrived, Saddam got in the back with Jajo to take him to the Ibn Sina Hospital, but according to Yahia the double, Jajo was already dead.

Uday had meanwhile swallowed a bottle of sleeping pills and was taken to the same hospital. As his stomach was being pumped out, Saddam arrived in the emergency room, pushed the doctors aside, and hit Uday in the face, shouting: "Your blood will flow like my friend's!"

Frantic at the possibility that Saddam might actually kill their son, Sajida turned to none other than King Hussein of Jordan, who was at the time apparently close enough to the Iraqi ruling house to function as a family counselor. Calling the royal palace in Amman, she shouted, "Uday has killed Jajo and now Saddam wants to shoot Uday." Without explaining to anyone what had happened, the king drove straight to the airport and flew to Baghdad. As he related later, he and the Iraqi ruling family spent the next several days "talking the whole thing over."

Whether because of the king's counseling, Sajida's pleadings, or simple paternal affection, Saddam's anger cooled. The killing was initially completely hushed up, with no word in the press until a month had passed. Then, on November 22, Saddam called publicly for the justice minister to investigate what had happened. He declared that Uday had killed Jajo unintentionally, adding that his son had been in prison for a month and had tried to kill himself three times. A government-orchestrated press campaign begged Saddam to show leniency. Ultimately, a three-man commission looked into the affair and freed Uday, who then left the country to stay with his uncle Barzan in Geneva. His behavior did not improve. Detained by the Swiss police for carrying a concealed weapon, he was asked to leave Switzerland.

In Baghdad, the family row over Saddam's mistress and the murder on the "Mother of Pigs" island had a final, mysterious chapter. Khairallah Tulfah, the Iraqi defense minister, took his sister Sajida's side in the affair and there were rumors in Baghdad that his support would cost him his job. In fact, he died in a helicopter crash the following year. The official story was that the crash was due to a sandstorm, but his father believed his son's death was too opportune and that Saddam had arranged for his murder by having explosive charges placed in the helicopter. Others suggest that for once Saddam was innocent because there was indeed a blinding sandstorm that day.

Even though Saddam brought the late defense minister's two little sons to live with him at the palace, Khairallah's entire family, as well as his father, had no difficulty in believing that he had been killed on Saddam's orders. (Saddam might have had an added incentive to get rid of his cousin, given the latter's popularity with the army and the ability he had shown as a commander in the war with Iran.) Yet they could not escape from the court of the ruler they believed was their brother's murderer. "Abdul" (not his real name), a young Iraqi businessman from an old family that moved abroad after the 1958 revolution, saw a lot of Uday and his set during visits to Baghdad at this time. "The whole Khairallah family did hold Saddam responsible for Adnan Khairallah's death. They all hated Saddam," he recalls. If they indeed felt that way, it was all the more macabre that the late defense minister's half-brothers, Luai, Ma'an, Muhvar, and Kahlan, were forced to spend much of their time in Uday's company. There was no refusing an invitation from the prince. Uday especially liked Luai, an affection that may have been based on their equally vicious characters. When Luai was in school, he had kidnapped a teacher who had displeased him and ordered his bodyguards to beat up the unfortunate educator. But even in the ruling family, only Uday and Qusay enjoyed total immunity from the law. When Saddam found out what had happened, he summoned Luai to a family gathering and broke his arm with a blow from a stick while a video camera recorded the administration of family justice.

"They are all animals," says Abdul of these characters with whom he socialized for a time. Nevertheless, there was a pecking order in this jungle. Abdul noticed how terrified all of the younger members of the family were of cousin Uday, "especially Ma'an, who is large and fat and slightly slow-witted. Uday liked to tease him. When Uday was in the room, Ma'an would sit silent, not daring to move." Uday had a high turnover of friends, demanding their constant company and attention before dropping them after a few months. "Friends who have been discarded are so happy," remembers Abdul.

Abdul himself was conscious of the dangers of proximity to Uday, but the Iraqi businessman endured the relationship because his family hoped to regain extensive holdings in Iraq that had been nationalized years before. Even so, he came to dread the constant phone calls to his Paris office beseeching him to come to Baghdad on the next plane. He explains the basis of his relationship with Uday as the envy of a self-consciously parochial neophyte for the sophisticated Westerner. "He was always very polite and hospitable. I think he looked up to me because I grew up in Europe, I had always lived in the West, and I was rich," he observes. "Of course, he was far richer than I was—he had all of Iraq. But that didn't seem to matter. He was always asking my opinion of things."

This reverence for Western taste and culture was coupled with a commensurate and limitless disdain for his own people. On one occasion in 1990, Abdul was driving with Uday in Baghdad when they passed two young boys, about ten years old, who were eating sunflower seeds and spitting out the shells. "He picked up the phone and called his guards in the car behind to go and beat up these boys because he thought they were spitting at him! 'You have no idea what *shits* Iraqis are,'" he expostulated as they drove on.

After his return from Switzerland, Uday's rehabilitation was swift. He was reelected as chairman of the Olympic committee from which he had resigned after the murder of Jajo. He wrote the introduction to a biography of his father. A few days before his father invaded Kuwait, Uday was part of the Iraqi delegation who went to Jidda on July 31, 1990, for a final disastrous meeting with the Kuwaitis. But his understanding of the world was still as parochial as ever. Abdul was with him on the day that the United Nations imposed sanctions on Iraq, four days after the invasion. Uday asked him how long he thought the sanctions would last. Abdul actually believed they would be in place for a long time, but to be on the safe side he said, "Maybe a month." Uday looked at him as if he were mad. "He said, 'You're joking—two or three days at the most. Do you really think the oil companies will let them go on longer than that!' You see," explained Abdul, "he and the rest of his family thought that Iraq was the center of the world, that the world could not live without Iraq and its oil."

That assumption, possibly shared by Saddam himself, had proven catastrophically wrong. Eighteen months later, the sanctions were still in place as Uday, his brother, and his cronies sang and talked at the National Restaurant in the al-Rashid Hotel. Perhaps it was the envious interest of Uday in the unreachable world of the

West that prompted his friend Ahmed to invite the American journalists lingering in the empty restaurant to meet the "cubs." It was
an irresistible invitation.

At the table, Uday was very clearly the dominant figure. Qusay,
dressed out of a J. Crew catalog, gave an impression of being the boy
at the dance with sweaty palms who would rather be dissecting rats at
home. Flushed with shyness, he spoke quietly about Mesopotamian
culture, occasionally glancing furtively at his elder brother as if asking
for approval. Only two years younger than Uday, he still appeared very
much the baby boy, but a polite inquiry about what he did elicited a
word of praise from his sibling. "Not such a baby now," laughed elder
brother indulgently. "He runs all the security services."

The outside world was as yet barely aware of the rising power of
mild-mannered Qusay in the regime's apparatus of repression,
although he was already incorporated into the family's personality
cult. One propaganda mural of the time showed Saddam as the
paternal sheikh on horseback with his rifle and flanked on each side
by both his sons, hawk-eyed defenders of their father. The younger
son had obviously inherited his father's skill in administration as well
as his habits of hard work. In the years to come, his executive
responsibilities would only grow, to encompass such vital tasks as
playing the different Kurdish factions off against each other, directing the battle of wits with the Unscom inspectors, and, above all,
guarding the personal security of his father.

Saddam, always a shrewd judge of managerial talent, was never
to grant Uday any similar official responsibility, but at the time of
this dinner, the elder son had been permitted to launch a new newspaper called *Babel*. It was a sign that Uday's power was expanding
well beyond the modest prerogatives of the Olympic committee.

"Iraq's only independent daily," as its editor proudly described it,
was already a smashing success and the talk of Baghdad because of
its uninhibited attacks on government officials (the father of the
proprietor excepted). It was powerful not just because it was the
voice of Uday but also because it was interesting and irreverent.
Recently it had run a competition for readers to vote on the most
handsome minister, with the health minister winning by a landslide.
Gossip about ministers, denunciations of black market racketeers,

exposés of shoddy work in the reconstruction of war damage—all were there in *Babel*. Not long before the dinner at the al-Rashid, the paper had revealed that the Jumhuriya Bridge across the Tigris, a cherished city landmark bombed and hastily rebuilt, was buckling in the center. "Our government provides us with everything, even an opposition," remarked Professor Jawad sourly. But the crusading paper was a shrewd initiative in furnishing an outlet for Iraqis traumatized by war, inflation, and a collapsing economy to vent their frustrations. A hundred or so citizens were turning up every day at *Babel*'s offices on Palestine Road with complaints about officialdom.

Sitting on a chair down the table from Uday in the restaurant, chain-smoking and nursing a whisky, was Abbas Jenabi, *Babel*'s editor-in-chief. Asked where he worked before *Babel*, he explained that he had been a foreign correspondent for the Iraqi News Agency, posted to Havana. He had not enjoyed his time in Cuba, he said, "because of the sheer lack of personal freedom."

The conversation turned to news leaking out of Washington about CIA covert operations under way to "get" Saddam. The "cubs" joined in the general hearty laughter at the notion, recently promulgated by a Pentagon spokesman, that there were "cracks in the inner circle" around Saddam, which could be exploited by giving greater support to the Kurdish and Shia opposition. "Look," said Uday, his enormous brown eyes narrowing, "the business about splitting us is nonsense." He gestured with his cigar around the table, which at this stage was littered with bottles of whisky, saying: "I have two Shiites here. I have a Kurd who works for me." Jenabi, a Shiite from a powerful tribe, nodded in tandem. "I have a big family," he said proudly. "There are two million of us." This Shiite, a loyal adherent of the Sunni ruling clique, was a case study in the shifting alliances of Iraqi politics. At an earlier stage of his career, he had worked on a paper owned by the Barzani family, perennial leaders of Kurdish rebellions. Later, in 1998, Jenabi fled Iraq, bearing vivid tales of the monstrous habits of his master, Uday and the torture he had suffered at his hands.

No one at this table appeared to have suffered ill effects from war or sanctions. "We spent the war at His Excellency's [meaning Uday] place in the country," remarked one of the party in a jovial aside. "Drinking and playing cards and watching the cruise missiles going overhead." Now this man was doing well in "import-export" with Turkey. A few years later, Saddam was to complain to Uday about the seediness of the company he kept. This was clearly evident that evening at the the al-Rashid, where everyone in the party was profiting in some way from their connection with the First Family. Down from Their Excellencies were two Armenians, one with an enormous gut who was boisterously drunk, the other very thin and even drunker. The thin one was the Saddam family jeweler, who was doing well out of the travails of ordinary Iraqis forced to pawn their jewelry to buy ever more expensive food. Haroot, his well-fed companion one place away, was Saddam's tailor, lauded by the rest of the company as a "philoso-pher." As the subject of conversation turned to America, Haroot passed a beefy hand over his pomaded silver hair and asked, "Did you know Howard Hughes? Well, you know who he was. You know the Armenian who works with him? Owns all the casinos?" When the answer was negative, he crowed with delight. "How come I know more about America than you do? Because I'm from the [Armenian] Mafia. If you ever need any help, the Mafia can help you. I'll fix it." He mused on happy days in Las Vegas. "What is the name of this guy, you know, with the rings and the big heart? Liberace! Oh, he was such a good man." Haroot turned to Saddam's eldest. "Do you know Liber-ace, Las Vegas?"

"No," said Uday with a lazy stare, "only Engelbert Humperdinck."

Given Uday's infamy, it would have been hard to forget, even in the midst of such a surreal conversation, that this was extremely dangerous company. Even so, his demeanor, together with that of his brother and associates, remained polite and courteous through-out. If a reminder of the true state of affairs was needed, it came in the form of the young security officer who came swaggering into the nearly empty restaurant. Suddenly, he registered the identity of the diners at the corner table. Within seconds he had turned on his heel and bolted out the door.

Uday was already using his newspapers to attack junior government ministers. Over the next few years, he would raise his sights to take on more formidable targets, ultimately clashing with senior members of his own family. But he was competing with other Iraqi political nota-bles for more than just political influence. For Uday, greed was just as important a motivating factor as power. From behind a cloud of cigar smoke at the head of the table, he took up the theme that there was an upside to the unfortunate fact that Iraq's economy was devastated by war and sanctions. "With all this situation, there is a lot of trade to be done, so I am doing some trade."

He was being too modest. The newspaper was merely the flag-ship of a growing business empire. He also controlled Babel TV, as well as Babel Transportation, Babel Hotels, and Babel Food Pro-cessing. When Iraq was exporting oil, members of the ruling clique could make great fortunes by taking commissions on contracts worth hundreds of millions of dollars to foreign companies. Now the government was poor, so Uday, Hussein Kamel, and the others were instead using their political and security muscle to acquire monopo-lies in importing consumer goods such as food and cigarettes as well as the profitable trade in smuggling oil in trucks across Kurdish-held territory.

Abdul's experience of Uday's approach to business, which he sum-marizes as "No money down and demand fifty percent," had left him unimpressed. His period of greatest contact with Uday was in 1990, prior to the invasion of Kuwait, at a time when Iraqi government pol-icy was to open up the economy to private business. But even with the backing of his powerful friend, Abdul failed to regain his family's for-mer businesses. "We didn't succeed," he explains, "because any indus-trial project or undertaking fell under the control of Hussein Kamel, and Kamel wanted the business for himself. For a long time, even Uday couldn't go up against Kamel in business."

That had been before the war. Following the disasters of 1991, as his father permitted him greater political rein, Uday used his new-found status to compete on more equal ground with the hitherto unassailable Kamel, who naturally resented this intrusion. In raising the profile of his eldest son, Saddam was exacerbating tensions within the ruling family. It was a potentially fatal development.

Nonetheless, Uday's power steadily increased, as did the status of his targets. By February 1994, *Babel* was attacking his uncle Wat-ban, interior minister at the time, for failing to prevent a series of ten "extremely successful" terrorist bombing attacks in Baghdad over the previous two years—twice as many, as *Babel* malignly pointed out, as had struck Baghdad during the entire eight years of the war with Iran. In March 1994, Uday was appointed leader of a new institution, "the Saddamists' Union." All high-ranking officials and army officers enrolled. Members carried special identity cards, giving them the right to benefits such as salary increases, special loans, and the right to a place at a university regardless of age or qualifications. It soon had twenty-five thousand members, most of them in the army. At the end of the year, the union spawned a special fifteen-thousand-strong militia known as the Firqat Fida'iyyi Saddam, or Saddam's Commandos, also led by Uday. He now had his own security force, an area that had previously been the preserve of his brother. The commandos, largely teenage toughs, often drove around Baghdad in pickup trucks with heavy machine guns in the backs, reminding visitors of the murderous militiamen of Lebanon and Somalia.

In the spring of 1994, Uday had assumed a further responsibility as overseer of the entire Iraqi media. From this command post, he promoted universal criticism of government executives for negligence and incompetence and ignoring the wisdom of "the comrade leader." In particular, he incited abuse against the hapless prime minister, a Baath Party veteran named Ahmed Hussein (unrelated to the ruling clan) for failing to prevent the further collapse of the currency.

The prime minister was duly dismissed in May 1994, and Saddam Hussein himself took over the post. Iraqis were informed that he would now do something to "dispel darkness and despair and create hope." The president/prime minister went out of his way to give an impression of activism. He announced that ministers who were not sick "should be in their ministries at 0:800. I want no other excuses."

Iraq was well past the stage when official punctuality made any difference. The economic decline was mirrored by the precipitate fall in the dinar, despite Saddam's more active role, from 140 to the dollar at the beginning of 1994 to 700 to the dollar in December. Baghdad looked like an enormous flea market as people sold off their household goods. The already meager monthly rations were reduced.

Uday meanwhile appeared to be going from strength to strength. His profitable business relationships with smugglers, as well as influential figures in Kurdistan, enhanced his relative position in "trade," while that of Hussein Kamel, so dominant only a few years before, appeared to decline.

Uday's elevation may have seemed a shrewd move to Saddam, always a deft manipulator in internal Iraqi politics. But in using his son to diminish other powerful members of the ruling family, Saddam ran the risk of tearing the ruling house, once notable for its unity, into fractious pieces. The use Uday made of his victory in gaining his father's favor was soon to provoke the worst political crisis since the days of the uprising.

In the meantime, in the rugged northern mountains of Iraq and far across the sea in Washington, D.C., potent enemies were pondering fresh attacks on the entire edifice of Saddam's rule.

SEVEN

Intrigue in the Mountains

On October 10, 1994, Columbus Day, a group of senior officials gathered in the White House Situation Room, the traditional venue for urgent and secret discussions on affairs of national security. Just a few days before, the U.S. government had learned that Saddam Hussein was moving troops toward the Kuwaiti border. Once again the Iraqi leader was demonstrating his ability to seize the headlines, disturb the weekends of senior officials, and prompt hurried movements of troops, aircraft, and warships.

The last time Saddam had moved toward Kuwait had been just before the 1990 invasion. On that occasion, George Bush had done nothing. Bill Clinton was naturally determined to avoid committing the same mistake, and so from the continental United States to the Indian Ocean, American forces were on the move. The President canceled a campaign trip to New Mexico and addressed the American people in martial tones. Saddam Hussein, Clinton declared, would not be allowed to "defy the will of the United States and the international community."

The meeting in the situation room was not concerned with the well-publicized movements of U.S. forces, but convened to review progress on a secret effort to eliminate the problem of Saddam Hussein once and for all. Back in April 1991 (see Chapter Two), the Bush administration had embarked on a two-track strategy toward Iraq: Saddam would be hemmed in by sanctions while the CIA simultaneously worked to bring him down. The Clinton administration had left the Bush approach essentially unchanged (apart from an offhand remark by the President-elect, hurriedly renounced, to the effect that normal relations with Saddam were possible). Sanctions were maintained as rigorously as ever. In 1993, Vice President Al Gore announced plans to seek a United Nations investigation of war crimes by the Iraqi regime, though nothing further was ever heard of the idea. When details emerged of a scheme by elements of the Iraqi security service, in association with a gang of whisky smugglers, to assassinate ex-president George Bush during a visit to Kuwait in 1993, Clinton fired off twenty-three cruise missiles at Iraqi intelligence headquarters in Baghdad, one of which went astray and killed Leilah Attar, Iraq's leading female artist.

In secret, Clinton reaffirmed Bush's directive to the CIA to unseat Saddam. Back in 1991, the agency, casting around for possible mechanisms to accomplish the task, had accepted the services of Ahmad Chalabi, the exiled Iraqi millionaire whose banking career had ended when his Jordanian bank collapsed amid charges of fraud and embezzlement. By the following year, Chalabi had become the moving spirit in an umbrella organization of opposition groups called the Iraqi National Congress (INC), pledged to overthrow Saddam and institute democracy in Iraq. Unbeknownst to most of those involved in the INC (apart from Chalabi) the organization's funding came from the CIA. Much of this money—over $23 million in the first year alone—was invested in an anti-Saddam propaganda campaign directed at audiences both inside and outside Iraq and partly designed to deflect international concern over the suffering caused by sanctions. This campaign was subcontracted to John Rendon, a Washington PR specialist with excellent agency connections.

While the connection between the agency and the INC was a closely held secret, the publicly expressed aims of the opposition coalition were perfectly respectable: a democratic Iraq with a government that would represent all races and creeds. The founding members had included individuals and groups from across the political spectrum of the Iraqi opposition. There were explicitly Islamic elements like the Shiite exile Mohammed Bahr al-Ulum as well as remnants of the once powerful Iraqi Communist Party, the Sunni ex-general and ambassador Hassan al-Naquib, and liberals like Laith Kubba, the civil engineer who had traveled around the United States in 1988 in a lonely effort to draw attention to the slaughter at Halabja. The largest and most important of the groups affiliated with the INC were the Kurdish parties, who were also the only organizations with significant military forces at their command.

Most of those who pledged support to this had spent the greater part of their lives in opposition to Saddam and Baathist rule. Some, however, had once been members in good standing of the Baathist regime before gravitating to the opposition. Many of this latter group, for the most part Sunnis, had found a home in a group called the Iraqi National Accord, or "al-Wifaq." The Accord had affiliated itself with the INC in 1992, but from the outset pursued its own agenda.

Chalabi and his colleagues believed that the way to undermine Saddam was from below, by sapping the dictator's power from a base in liberated Kurdistan through such means as propaganda and the encouragement of defections by officials of the regime and desertions among the army. As endorsed by the CIA, this was an essentially political operation. So long as the INC confined itself to the role of democratic opposition to Saddam and promoting disaffection among the populace, the paymasters were happy. More aggressive initiatives on the part of the INC were less warmly received in Washington.

Even so, despite the fact that he was on the agency's payroll, Chalabi was not shy about promoting his views and agenda. In November 1993, he flew to Washington to unveil an ambitious plan to foment mutinies in army units around Iraq, which would eventually spread to Baghdad and topple Saddam. Addressing officials from the CIA, the State Department, and the Pentagon at the Key Bridge Marriott Hotel (a favored watering hole of the intelligence community), he gave precise details of the adventurous scheme and outlined the support he would need from the United States to carry it out. Then he flew home and waited for a response. There was none.

The problem with Chalabi's grand plan, so far as Washington was concerned, was precisely the question of American support. Any Iraqi unit that defected en masse would certainly evoke a violent response from Saddam. However decayed and disaffected the bulk of the Iraqi army might be, he could still count on the comparatively well-armed Republican Guard. Resisting the counterattack, therefore, would require help from the Pentagon in the form of air support. But the U.S. military was very dubious about involving itself in fighting in Iraq.

"I would go to the JCS [Joint Chiefs of Staff] and say, 'If I can identify a military unit that is ready to mutiny, will you adopt it?'" recalls one CIA official involved in the Iraq operation. "The answer was never no. The answer was never yes. It was always 'We'll get back to you.'"

Opposition to Chalabi's grand initiative was not confined to the U.S. military. According to Frank Anderson, then head of the Near East Division (colloquially referred to at Langley as NE) of the agency's covert operations directorate, he thought at the time that the INC represented merely "the capability to be another problem for Saddam, in fact, a serious problem," but nothing more. In their hearts, the decision makers yearned for the simple solution, a palace coup that would replace Saddam with a (hopefully) more benign and well-disposed strongman. Ever since President Bush had issued his finding in May 1991, the agency had been attentively waiting for a person or persons who might mount such a coup. In the summer of 1994, hope began to burgeon in the breasts of some agency officials that deliverance might be at hand.

It was not an entirely unrealistic proposition. As we have seen, there had been a series of conspiracies against Saddam by officers from Sunni Muslim tribes, like the Juburi and the Dalaim, that traditionally supported the regime. All were detected and the conspirators ruthlessly punished, but if the agency could only make contact with the right group in time, then a successful coup might be possible. Unfortunately, almost the only points of access for the CIA to internal Iraqi dissent were through exile groups, themselves under the unblinking scrutiny of Saddam's intelligence services.

Iyad Alawi, the leader of the Accord, was a charming and articulate individual who had the gift of impressing intelligence officials. He had long nurtured close links with MI-6, British intelligence, who cherished him as an old and valued agent. "The Brits are always falling in love with people," recalls one CIA officer involved in the

Iraq operation of his transatlantic colleagues. "They are romantic in that way. Funnily enough, the FBI are the same."

From early in 1994 on, Alawi began a series of intense meetings with his British friends in London and various vacation resorts on the south coast of England. The news he reported from his contacts in Iraq appeared to hold out exciting prospects of unrest at senior levels in the Iraqi army. All that was needed, he reminded his eager interlocutors, was support and, above all, money from outside. MI-6 passed on the news to the "cousins" in the CIA London station, who in turn urged the merits of the INA to sympathetic ears back at Langley.

This, then, was the background to the gathering at the White House on Columbus Day, 1994. The senior officials assembled in the situation room included Peter Tarnoff, the undersecretary of state for political affairs; George Tenet, the National Security Council director for intelligence affairs; Madeleine Albright, U.S. ambassador to the United Nations; and Admiral David Jeremiah, chairman of the Joint Chiefs of Staff. They had commissioned this classified briefing to hear what the agency was really accomplishing behind enemy lines.

Leading the CIA delegation was the deputy director for Operations, the man in charge of all CIA covert actions, Ted Price. A Yale graduate and ex-marine, Price was a man who easily impressed people. Like many officials who rose to high rank in the operations directorate in the 1980s, he was a graduate of what were called the "hard language" programs—Arabic and Chinese. Price's speciality was China and he spoke fluent Mandarin. No one disputed that he was highly intelligent, though one former colleague describes him as "quick but not wise." He was also intensely ambitious, having set his sights on the coveted position of "DO" long before he was appointed in December 1993. A former chief of another U.S. intelligence agency recalls the short, sandy-haired Price as being "very smart, as much a politician as a professional." Given, therefore, the long-standing preference in the White House for a "silver bullet" coup by members of the Sunni power elite, which would replace Saddam without totally upsetting the Iraqi political order, it was natural for Price to tout the CIA's potent contacts in such circles.

The centerpiece of the briefing, as planned by Price, was a chart depicting the agency's network inside Saddam's regime. Tightly packed with the names of officials from the Iraqi military, intelligence, and other key components of the Iraqi ruling apparatus, it certainly made for a striking display. "The way the names were depicted on the chart," recalls one official familiar with the presentation, "it looked like he [Saddam] was surrounded."

However, there were those back at CIA headquarters in Langley who thought that perhaps this tableau of Iraqi dissidence gave a far greater impression of CIA connections inside the country than was, in fact, the case. "If you say we have a direct communication from so-and-so," recalls this same official, "that's different from someone crossing the border with a note saying 'My cousin Ali hates Saddam and wants to bring him down.' A lot of the names on the chart had been generated in the second way."

Price was not the only important figure from the agency among those present. By his side was Frank Anderson, the man who had been ordered by George Bush to "create the conditions" for the removal of Saddam Hussein nearly three and a half years before. Since the day that he scribbled "I don't like this" on Bush's directive, Anderson's pessimism about the prospects for bringing Saddam down by any kind of covert action had not lifted. In fact, according to others involved in the operation, he had paid as little attention as possible to the day-to-day details. "Frank would help out if, for example, I was having a problem with the State Department, but for the most part he was far more involved in the Israeli-Palestinian peace deal that was happening at that time," recalls one of his former subordinates. "He certainly never turned up at any briefings on Iraq if he could help it."

The Columbus Day briefing was, apparently, an exception. If Price was determined to put on a big show, Anderson had every reason to turn up. Relations between the two men were not good.

Ever since the arrest in February 1994 of Aldrich Ames, the Russian spy in the heart of the Operations directorate, CIA director James Woolsey had been under heavy pressure to fire senior officials who had failed to take note of the alcoholic Ames as he flaunted his earnings from Moscow for betraying most of the agency's spies in Russia. One of these officials was Ted Price who, in his former post as head of counterintelligence, had been oblivious to the mole under his nose. Price was not fired but merely reprimanded and remained as deputy director for Operations. Woolsey did, however, decree that no one who had ever been Ames's superior should be given any sort of agency award or commendation.

Frank Anderson had never been in contact with Ames, but his old friend Milt Bearden had. Bearden was best known inside the agency for his impressive work in the Near East Division master-minding the shipment of huge quantities of weapons and money to the Afghan Mujaheddin for their war with the occupying Soviets in the 1980s. Following that triumph, Bearden had taken over the Soviet division, where Ames worked. Two weeks before the White House briefing on Iraq, Bearden had retired.

Despite Woolsey's edict on awards, Anderson and another senior operations veteran and friend of Bearden, John MacGaffin, had decided that their old comrade in arms should not be allowed to disappear into retirement without some small recognition of his Afghan triumphs. Anderson had therefore presented Bearden with a plaque from his colleagues. Word of this "transgression" sped back to Langley and the receptive ears of Ted Price, who hastened to apprise Woolsey of what had happened.

Woolsey, a lawyer and a defense intellectual, rather liked derring-do, cloak-and-dagger types, but following Price's report he felt he had little option other than punishing Anderson and MacGaffin by demoting them. They denied him that option by quitting. Thus, when Anderson arrived at the White House for the briefing, he knew his career was over and who was responsible.

He could have left the briefing to Price, the most senior agency official present. Instead, he stepped up beside the chart and proceeded to tear to pieces the entire elaborately crafted presentation of CIA prowess. Anderson pointed out that the carefully delineated lines of communication between Saddam's security apparatus and the agency stations in neighboring countries were little more than rumors, and that those who might have sent messages indicating a willingness to conspire against the leader were just as likely to be double agents ultimately controlled by the spymasters in Baghdad.

It was a withering performance. The high-level group in the room, who had arrived in the expectation of good news, was aghast and indignant. After Anderson had finished, Madeleine Albright spoke up in exasperation. "Why are we here?" she asked angrily.

Anderson, as one CIA official discussing the events of the day put it, had "rained on the parade." But this was his swan song. It was not that he dismissed the idea of pursuing the possibilities of a coup, or favored one approach to the Iraqi problem over another. He had simply decided to give the policy makers the unvarnished truth; casting a blight on Price's big day may have been an added incentive. However, the higher-level bureaucracy is congenitally unwilling to accept bad news that conflicts with its deep-set hopes and desires. Anderson had suggested that the glass was at least half empty. Despite his performance, officials at the highest levels persisted in believing it was half full. In the White House itself, for example, the director for intelligence affairs on the National Security Council staff, George Tenet, was a powerful advocate of the coup option. His boss, National Security adviser Tony Lake, was no less intrigued by the possibility.

As head of NE, Frank Anderson had been catholic in his attitude toward all aspects of the operation against Saddam. Lacking much faith that anything was likely to work, he had been happy to endorse all avenues. If the London station was excited about Iyad Alawi and his plans for a coup, Anderson was happy to let them proceed. On the other hand, there was the ongoing effort with Ahmad Chalabi and the Iraqi National Congress. Anderson was happy for that to proceed also. A month before his climactic interview with Jim Woolsey, Anderson had given the go-ahead to send a team of CIA officers into Iraq. They would work with the INC at their base in the liberated zone of Iraqi Kurdistan.

"What I wanted them to do," as Anderson later explained his decision, "was to be in a position to look for this hoped-for, but yet-to-be-achieved coalition of forces that might put us in a position to move forward against the Saddam regime." Anderson gave a wry smile as he said this, a possible indication that he personally had little faith in the prospects of any such forces appearing.

The decision to send CIA officers into Iraq had powerful support on Capitol Hill. In September 1994, two members of the Senate Intelligence Committee staff, Chris Straub and Don Mitchell, had ventured into Kurdistan on a fact-finding mission and had met with

Ahmad Chalabi, the Kurdish leader Massoud Barzani, and the former general Hassan al-Naquib, customarily trotted out on such occasions as a representative of support for the INC in senior Sunni military circles. As a result of their encounter with these doughty freedom fighters, Straub and Mitchell returned to Washington vastly impressed. Senate support for the agency's clients in northern Iraq increased commensurately and the dispatch of a team of actual CIA officers was speedily approved by the Intelligence Committee. In early October, the first team, led by a ruddy-faced Chicagoan named Warren Marik, arrived to set up shop.

The four men on this team, like those who followed them, were field officers, the agency equivalents of lieutenant colonels and majors, far removed in rank and power from senior officials like Anderson and Price. Policy was conceived and argued over high above their pay grade. Their job was to deal with the surrogates, in this case Kurds and opposition Iraqis, in order to collect and evaluate the information seeping out of Iraq. They had done this kind of thing before. Marik, for example, had joined the CIA after fighting with the U.S. Army in Vietnam, and had then been part of the massive agency operation training and supporting the Afghan Mujaheddin against the Russians in the 1980s. For a year before he went to Kurdistan, he had been assigned to the Iraq office at CIA headquarters in Langley, acting under "Big Ron" Wren, the man in immediate charge of the Iraq operation. While there he had dealt with administrative tasks, including fruitless attempts to rein in the enormous costs of John Rendon's propaganda operation. "Every time something happened in Iraq, John would jump on the Concorde," he lamented later. While at headquarters, he had been in almost daily contact with Chalabi, far away in Kurdistan. Now he and his colleagues were going into the heart of things.

Salahudin, where the Americans were housed in a heavily guarded villa, sits high up on the western fringe of the dramatically beautiful Zagros Mountains, which stretch across Kurdistan and into Iran. The villa looked out over the plains that extend all the way to the Persian Gulf, far to the south, and in the near distance, the city of Arbil, a forty-five-minute drive down the switchback road that leads to the plains. In the 1970s, it had been developed as a summer resort, complete with some indifferent hotels and prefabricated Swiss-style chalets where middle-class families from Arbil would go to escape the scorching summer heat. By 1994, the vacationers were long gone and the hotels were occupied by the Iraqi opposition. The INC had taken over an entire hotel, decorating it with lurid posters depicting the imminent defeat of Saddam. Chalabi and his staff had all rented houses for themselves. There were also offices to run the INC's radio and TV services and one that produced its newspaper. "It was like a mini-state," fondly recalls one INC activist of those days. Since the INC was funded by the CIA, this mini-state constituted an agency operation comparable in scale to the infamous Bay of Pigs effort against Castro almost thirty years before. When disaster struck the INC two years later, as many as five thousand people had to be evacuated to safety by the United States.

Despite the fact that the Kurdish enclave was officially sealed off from the rest of Iraq, there was, nonetheless, considerable communication between the two regions. Individuals, even Iraqi army officers, went back and forth to visit friends and relatives. An extensive smuggling network crossed the lines, hauling diesel fuel up from the oil fields (still firmly in Saddam's hands) to the border crossing with Turkey at Khabur, controlled by the Kurds, and food and other consumer goods back down. Using such routes, the INC (like the Kurdish groups and the host of foreign intelligence agencies at work in the north) was able to establish its own network of contacts to relay news of what was going on behind the lines. The mechanism of this arrangement was fueled by money, ultimately supplied by the CIA. Unfortunately, the nature of the system encouraged production but not always accuracy, since those who had information to supply got paid, while those who had no news did not.

A related "product" of the INC intelligence system was the individuals fleeing Saddam's regime. Many of these people had occupied sensitive positions before leaving. Those who had valuable secrets or contacts to divulge could hope to be passed on into the greater freedom of the outside world, with an American green card as the ultimate prize. Sometimes these arrivals were of great importance—high-ranking generals or officers from the intelligence services. Others were more problematic, minor figures eager to inflate their importance in the Baghdad regime in the hope of a speedy passage to the United States.

Professional intelligence officers on the spot to debrief such "walk-ins" would minimize the risk of missing important information, even if all interrogation had to be carried out through interpreters (supplied by Chalabi or one of the Kurdish groups). Only one of the CIA officers posted to Salahudin between October 1994 and March 1995 spoke Arabic; none spoke Kurdish. Marik himself could get along in Turkish and a few of his colleagues had learned Farsi, the language of Iran.

For the INC, the presence of the Americans could be presented as an affirmation of American support. Later, Chalabi claimed that the Americans had announced on their arrival that "the United States government has decided to get rid of Saddam Hussein, and we want your help on this." This was a rather more dramatic take on the purposes of the mission than the modest objectives in Anderson's mind when he issued the orders. The INC, with its feeble military strength and well-advertised antipathy to installing another Baathist military strongman in place of Saddam, were not considered a promising vehicle to "get rid" of Saddam. But the Americans were not there just to work with the INC. An especially secret component of their orders directed them to work directly with the Accord, bypassing the INC with which the Accord was nominally affiliated.

The Accord had its own presence in Kurdistan, headed by a former general in the Iraqi army named Adnan Nuri. Nuri, a somewhat sinister-looking Turcoman (a minority people in Iraq), had a direct connection with the CIA. In June 1992, just after the (CIA-funded) INC founding conference in Vienna, the agency had flown Nuri to Washington for a discreet meeting at the Sheraton Premiere Hotel in the Tyson's II suburban shopping mall. At the meeting, the agency representatives told him, as he later reported, "You work separately from the INC, but don't resign from the INC. Be in the INC, but work separately." The "work" the Americans had in mind was to facilitate a coup. In public, meanwhile, the U.S. government gave a very different impression of its plans for Iraq. Not long after the recruitment of Nuri, the White House assured the INC of its lack of interest in the "dictator option," i.e., a military coup to replace Saddam with another strongman.

Thus the American operation in Kurdistan was mired in intrigue and double-dealing from the moment the first team made its very visible appearance in Salahudin. ("Are you here as an overt or a covert operation?" maliciously inquired one Kurdish official.) Adding to the complexities of the CIA officers' situation was the fact that they were sitting in the middle of a political earthquake zone.

Ever since George Bush had been reluctantly impelled to send American troops into Kurdistan in April 1991, the Kurds had enjoyed a safe haven in their mountainous homeland. The troops had left, but daily flights of U.S. warplanes from their Incerlik base across the border in Turkey were a guarantee that Saddam's forces down on the plains would not come north again. As a result, the Iraqi Kurds were enjoying a de facto independence. Turkish Kurds under the banner of the militarily efficient and ruthless PKK guerrilla organization had been waging a bloody insurgency against Ankara ever since 1984, but while the death toll mounted, they seemed no nearer to achieving their goal. To the east, a short-lived bid for autonomy by the Kurds in Iran had been summarily crushed by the ayatollahs in Tehran soon after the Iranian revolution in 1979. Only in Iraq were these long-suffering people in a position to govern themselves. In the spring of 1992, they had held an election, voting for an assembly that provided the mandate for a Kurdish regional government. Launched with great enthusiasm, the Kurdish Regional Government (KRG) struggled manfully to cope with the overwhelming problems of a country with a minimally functioning economy devastated by war and seeded with land mines. The Kurds' problems were compounded by the fact that late in 1991, Saddam instituted his own sanctions on the Kurdish area, banning trade and refusing to allow the transport, via Baghdad, of humanitarian supplies by international agencies. Since the sanctions imposed by the United Nations made no distinction between liberated Kurdish Iraq and those parts of the country ruled by Saddam, the Kurds were under a double siege.

These pressing difficulties were caused by the recent wars and the rebellion that had devastated the region. In addition, the quasi-independent enclave had to contend with an equally destructive historical legacy.

Traditionally, the Kurds had always been cursed by rivalries and factionalism. Kurdish society had remained divided along tribal lines long after the societies around them had evolved into nation-states. Even as a sense of Kurdish nationalism began to emerge well into the twentieth century, such divisions remained a fatal weakness. Time and again a Kurdish leader challenging a central government in Baghdad would find that other leaders from rival tribes would seize the opportunity to make deals with the enemy, at the expense of the insurgent, in exchange for cash or increased local influence. It was a semifeudal society in which division had long been the natural order.

Adding to the difficulties of the Kurds, who numbered some 25 million, was the fact that they were divided by national borders. Concentrated in Turkey, Iraq, and Iran (with some in Syria and some in the former Soviet Union), they were strong enough to frighten and cause trouble for the central authorities in each of these countries but never strong or united enough to wrest control of their own homeland. One or another of these powers might, for their own reasons, give aid and encouragement to a neighbor's Kurds in one of their periodic rebellions, but never enough to ensure success and almost always with a view to ultimate betrayal. In 1974–75, for example, the Kurdish tribal leader Mustafa Barzani led a massive insurgency against the Baghdad government. He accepted military help from the shah of Iran, who, however, was only giving it as a means to pressure Baghdad into concessions in another area. Barzani foolishly believed that the involvement of the CIA on his side, on orders from the White House, was a guarantee against abandonment by his ally in Tehran. Once Saddam Hussein agreed to Iran's demands for increased control over the Shatt al-Arab waterway on the two countries' southern border, the shah abandoned Barzani without any protest from the U.S. government. As a subsequent U.S. congressional report noted, secret CIA documents on the operation clearly showed that the White House and the shah did not want their Kurdish allies to win. "They preferred instead that the insurgents simply continue a level of hostilities sufficient to sap the resources of [Iraq]. . . . Even in the context of covert action," said the report, "ours was a cynical exercise."

As shocking as the evidence of this perfidy may have been to decent Americans, such a doublecross by governments who used the Kurds for their own purposes was hardly novel. As Saddam Hussein himself took care to mention in his address to the nation halfway through the Kurdish uprising that followed the Gulf War: "Every Kurdish movement that was linked to the foreigner or relied on him politically, militarily, or materially brought only loss and destruction to our Kurdish people."

Once the Barzani uprising had been crushed, in 1975, the Kurdistan Democratic Party (KDP), the party that the old rebel had led to defeat, formally split apart. An urban intellectual, Jalal Talabani, who had long resented and contested the feudal control of the Barzani family, founded a rival group that he called the Patriotic Union of Kurdistan (PUK). Within a few years, the two sides were fighting bitterly among themselves. The KDP, now led by Massoud Barzani, a son of "Mullah Mustafa" (who had died in exile), accepted arms and money from Iran in exchange for help against Saddam when the Iran-Iraq war broke out in 1980. Talabani's PUK in turn formed a temporary alliance with Saddam. By 1986, the wheel had turned again. The Kurdish groups united in alliance, both accepting support from Tehran to fight Saddam—even as the Iraqi leader was busily subsidizing the Iranian Kurds so they would fight on his side.

As we have seen, the two main Kurdish leaders and their respective followers combined to plan another uprising in March 1991, although the rebellion that did break out was largely a spontaneous affair that swiftly ran out of control. Following the ejection of Saddam's forces from northern Iraq (with U.S. and allied help), and despite the fact that they had a joint interest in maintaining the fragile statelet of Iraqi Kurdistan, Barzani and Talabani were hardly united. The "government" was delegated to underlings while the two leaders concentrated on advancing their own interests. "They are obsessed with their party rivalry," one Kurdish politician told David MacDowall, the leading authority on the modern history of the Kurds. "They do not work out a common strategy. There is no strategy at all, except to get ahead of the other party." Each of them jockeyed for support from outside powers. Talabani made overtures to the Turks, while for a long period Barzani enjoyed the patronage of the Iranians. They both lobbied for the support of the most important patron of all: Washington. More covertly, they both maintained lines of communication with their old enemy in Baghdad.

However, while officially partners in governing Iraqi Kurdistan, the two Kurdish leaders were also leading members of the Iraqi National Congress. One of the reasons that the United States had warmed to the INC during its genesis was the fact that the Kurds had agreed to join, signifying that they were forswearing thoughts of independence—which would have deeply offended and alarmed Washington's Turkish ally—and pledging to remain part of a unified post-Saddam Iraq. In consequence, the INC gained a secure base inside Iraq's borders. It also gained, at least in theory, the potential use of the Kurdish groups' thirty thousand or so Peshmerga fighters. The INC did begin fielding its own force of a few hundred lightly armed troops in 1993, mostly deserters from the Iraqi army, and for outside consumption their numbers were inflated to "thousands."

The INC and its CIA backers therefore were dependent for their continued operation on political stability in a land where the natural order was dissension and fighting. If the Kurds were to escalate their arguments and intrigues and revert to actual warfare, then Saddam would have the opportunity to move back into the mountains. In that event, the INC would be doomed. In 1994, the Kurds began fighting again.

The immediate cause was money. Kurdistan, economically isolated and devastated by war, did possess one significant asset: the border between Iraq and Turkey. The lines of huge trucks, laden with sanctions-busting Iraqi diesel-fuel exports, waiting to cross into Turkey at Khabur, outside Zakho, provided a fruitful source of revenue to whoever controlled the crossing point. Every truck, coming or going, paid tolls, adding up to hundreds of millions of dollars a year. As it so happened, Zakho and its environs lay in the territory controlled by Massoud Barzani's KDP and was more immediately dominated by Barzani's nephew, Nachirvan. U.S. State Department officials involved in humanitarian aid took to calling him "the best-dressed man in Kurdistan," his wardrobe periodically updated by trips to Neiman Marcus and other fashionable stores in the malls around Washington.

Jalal Talabani's principal support lay farther to the east. While the

PUK controlled major cities such as Sulaimaniya as well as a few minor border crossings with Iran, Talabani did not directly dominate any lucrative border-crossing point. For a few years after the establishment of the Kurdish Regional Government in 1992, this was not a major bone of contention. The "minister of finance" in the government was a PUK man who was in charge of collecting the tolls at Khabur, closely watched by an emissary from the KDP. At least a proportion of the money found its way into the common kitty. Then, in May 1994, the Kurdish modus vivendi began to break down.

The immediate cause was a local land dispute between two groups, each allied with a different dominant faction. Neither of the two leaders could completely control his followers, and fighting gradually flared across the north, with casualties in the hundreds.

It fell to the INC leader Ahmad Chalabi to broker a cease-fire. Americans who were on the ground agreed that he and his subordinates—mostly Arabs, with some Kurds—were extraordinarily successful in the mediation effort. Essentially, the INC inserted itself between the two sides, sometimes when they were actually firing at each other, setting up checkpoints complete with flags on roads leading from one group's area to the other. By the end of August 1994, these unremitting efforts had paid off and an uneasy peace had descended over the mountains. As a result, the local prestige of the INC skyrocketed.

Chalabi, however, soon began complaining to Washington that keeping the peace in Kurdistan was expensive. In messages to headquarters, he cited the cost of maintaining checkpoints and teams ready to mediate whenever tension flared. INC personnel were well paid, at least as compared with the Peshmerga, who were paid little, when at all. He needed more money—a million dollars, he said—but all he received from Washington were exhortations to continue keeping the peace coupled with vague promises of payment at some later date. It seemed that Langley was losing interest in its protégés in Salahudin.

Warren Marik had every sympathy with Chalabi's predicament. In exasperated cables back to headquarters, he pointed out the importance of keeping the rival Kurds from each other's throats and the small amount of money involved. "All I got were sort of 'check's-in-the-mail-type promises,'" he said later. He was informed that there were legal problems in sending the money on the grounds that "there's no authorization to fund an INC mediation force." Only later did Marik come to understand that Langley's disinterest reflected the waning appeal of the INC in Washington.

In December 1995, the Kurds started fighting again, this time with greater intensity. Whatever revenues the PUK had been receiving indirectly from the border tolls at Khabur were now gone forever, as were Talabani's own highly profitable investments in freight companies at the border crossing with Turkey. The PUK leader, however, scored a considerable military success at Christmas by capturing the Kurdish "capital" of Arbil, down the mountain from Salahudin, ejecting the forces of the KDP. Kurdistan, where the Kurds had managed to rule themselves in freedom from Baghdad, was embroiled in a vicious civil war.

Just as the civil war broke out anew, a defector of extraordinary importance arrived in Salahudin. General Wafiq al-Samarrai had been preparing his escape from Baghdad for some time. In the summer of 1991, feeling that the Kurds had to come to some kind of settlement with their mortal enemy, since none of their neighbors would countenance an independent Kurdistan, Massoud Barzani had gone to Baghdad to negotiate. Since Kurdish affairs were traditionally the responsibility of military intelligence, al-Samarrai had been ordered by Saddam to escort the Kurdish delegation. During a meeting with Hussein Kamel, the Kurds watched, open mouthed, as the loutish but all-powerful Kamel cursed and abused al-Samarrai for some imagined infraction. In the car after the meeting, the Kurds asked al-Samarrai how he, as a high-ranking professional army officer, could tolerate such behavior from "a sergeant." Al-Samarrai gazed out the window and then muttered, "It might not always be like this."

That was dangerous talk in Saddam's Baghdad and both al-Samarrai and the Kurds knew it. When Barzani and his men returned to the north after breaking off negotiations with Saddam, the intelligence general stayed in touch. This was a hair-raisingly risky course of action, but, as a member of the Iraqi opposition who came to know the general well put it, "Wafiq was one of them, he knew how they operated their intelligence and how to evade them."

Early in 1992, Saddam transferred al-Samarrai to an intelligence post in the presidential palace, where he stayed until the day he was informed by a friend, late in 1994, that his suspicious master was planning to kill him. He was able to get a message to Barzani, asking for help in being smuggled out. The Kurds obliged, and on December 2, 1994, he walked into Salahudin. Some of the old friends he met there were very surprised indeed to see him. A newly arrived CIA officer took one look at the general and shouted delightedly: "Ali! What are you doing here?" It transpired that in a previous age, the officer had been posted to Baghdad as part of the liaison group sent by the agency to assist Saddam Hussein in his war with Iran. He had dealt directly with al-Samarrai, but had known him by the code name "Ali."

The assorted guerrillas and revolutionaries in the room watched bemusedly as the two intelligence professionals reminisced about old times. "It proves a point," laughs Ahmad Chalabi. "The only real friends and contacts the CIA had in Iraq were Baathists!"

The Iraqi intelligence general had many secrets to tell. For Unscom, he had news of Saddam's ongoing biological weapons program. For Chalabi, he had exciting news about conditions inside the Iraqi army and which commanders were disaffected. He mentioned that Saddam might be contemplating a visit to the city of Samarra, al-Samarrai's hometown, in the near future. Were that to happen, the general thought that it might be possible to enlist members of his own numerous and powerful clan to ambush the leader's cavalcade as it crossed a bridge into the city.

This kind of idle talk did not constitute a concrete assassination plot, and none of the seasoned conspirators who talked to him at that time took him particularly seriously. "Maybe Wafiq was exaggerating everything in order to give himself a big role in Kurdistan," suggested one senior INC official afterward. But discussion of the Samarra assassination scheme was to cause problems later on.

Chalabi, beset with money problems, an attitude of apparent indifference back in Langley, and bitter fighting between his Kurdish allies, saw a way of raising the profile of the INC with what he called the "two cities" plan. To the south of the liberated zone lay two large and important cities, Mosul and Kirkuk. If Saddam were to lose them, it would be a serious blow to his regime. The INC leader's notion was to apply the "carrot and stick." If the generals commanding the Iraqi army garrisons in and around the cities could be suborned to the extent that they would allow the INC free rein to infiltrate their cities, this would provoke a reaction from Saddam, who would order the generals to be relieved. It would then be put to these generals that rather than return to Baghdad to face their master's wrath, they should defect with their families to the INC. Should the generals resist these blandishments, then the "stick" would be applied in the form of military attacks by the INC and its allies. Given the poor shape of the ordinary Iraqi army units up near the front lines, these attacks would probably be successful enough to embarrass the generals and get them into trouble with Baghdad. Either way, the Iraqi military would be progressively weakened and even further demoralized, leading in turn to a progressive loss of Saddam's control. Al-Samarrai gave vocal encouragement to the scheme. According to some reports, the general was promising an uprising by his friends and supporters in military units around Iraq, to occur as soon as Saddam was distracted by an INC offensive in the north. In the best of all possible outcomes for Chalabi, an Iraqi counter-attack in the North would in turn prompt U.S. military intervention.

It was an intricate and imaginative plan, which any guerrilla commander operating with a secure base and firm support from outside might have approved as a sensible initiative in wearing down the enemy's main forces. But Chalabi did not have a secure base—his allies were fighting. Nor did he have firm support from outside, since the high command of the CIA was increasingly transfixed by the possibilities of a coup. He was running short of money and, in fact, had recently borrowed considerable sums from local businessmen grown rich in the smuggling trade. If, however, he were to score a dramatic success, the situation might be recouped. But he would have to act soon, since he knew full well that he had rivals for the agency's affections.

The senior management of the CIA's Near East Division might have been deluding themselves into thinking that the INA's preparations for a coup were cloaked in darkest secrecy, but they were wrong. "The INC knew exactly what was going on," recalls one of the field officers formerly stationed in northern Iraq. "The INA was as leaky as a sieve. Chalabi wanted to preempt them."

The catalyst that was to spark Chalabi's move—and in doing so ruin the prospects of the INC—arrived in Salahudin in early January 1995. "Bob," as he later became known in the media, was a lanky six-footer who had served in Afghanistan. In addition he knew the Middle East well from postings to various CIA stations in U.S. embassies around the region, including Syria. He spoke passable Arabic.

Bob's mission, as conceived at headquarters in Langley, appears to have been no different from that of the other agency officers who had been rotating in and out of Kurdistan since October. This was to collect intelligence as well as assist the purportedly super-secret plots of General Nuri and the Iraqi National Accord. Bob had not been in the country long when, it seems, he decided to try something more adventurous—a direct attack on the Iraqi regime.

For some of the characters who encountered this CIA officer at that time, his heady enthusiasm for energetic action against Saddam appeared ludicrously naive. To a man, the veterans of countless bitter encounters on the battlefield and at the bargaining table, betrayals, defeats, massacres, exile, shifting alliances, and other routine aspects of Iraqi Kurdish politics found Bob totally lacking in an understanding of the local situation. "I liked Bob," recalled Hoshyar Zibari, a senior adviser to Massoud Barzani. "He was a really interesting guy, fascinating to talk to about all sorts of things. But he had some *really* weird ideas."

The spark that impelled the CIA officer into full-scale action was struck far to the south of Kurdistan, in a place called al-Qurna, near the marshes that straddle the border between Iran and southern Iraq. On February 12, 1995, the 426th Brigade of the Iraqi army suffered a severe defeat in an armed clash there and lost several hundred men killed or taken prisoner. The attackers were part of the "Badr Brigade," Iraqi Shiites armed and financed by Iran.

It may be recalled that Iran had done little or nothing to help the Shiite rebels in March 1991. Despite fears in Washington that the rebellion was backed by Iran, the Tehran government, fearful in its turn of provoking the United States, had permitted only small groups of armed exiles to cross the border to help the uprising. The few that did were members of the same Badr Brigade, which had originally been raised by the Iraqi Shiite leader Mohammed Baqir al-Hakim and had fought with considerable effect on the Iranian side in the Iran-Iraq war. Since the uprising, the brigade had been permitted by the Iranian government to make armed incursions into southern Iraq, operating out of the sanctuary of the marshes. (These marshes were, however, a diminishing sanctuary, since Saddam was dealing with the problem by draining them, displacing the unique community of the Marsh Arabs who had lived there for thousands of years. One of the few people who worked to draw attention to their plight was Hussain al-Shahristani, now in exile in Iran.)

As Iraq's most powerful neighbor and bitter enemy, Iran was a crucial factor in the politics of the region, which the United States, as a matter of policy, was determined to ignore. In the summer of 1992, some officials in the National Security Council had made overtures to emissaries of Hakim with a view to assisting the southern fighters, embarrassing Saddam, and aiding President Bush's reelection campaign. The initiative foundered, though Bush did institute a southern "no-fly zone" for the Iraqis, enforced by allied aircraft, as an entirely symbolic gesture to aid those fighting in the south. Soon after taking office, the Clinton administration proclaimed the policy of "dual containment," under which Iraq and Iran were to be treated as pariahs of equal status. Cooperation with Iran against Saddam was absolutely out of the question. With the eager encouragement of Ahmad Chalabi and Wafiq al-Samarrai, Bob resolved to change all that.

"Bob got very excited when he heard what the Badr Brigade had done," says Chalabi. "He almost went berserk. It showed that the Iraqi army was vulnerable." The American and the Iraqi had already had long discussions about the "two cities" plan. Now, explained Chalabi later, "Bob said, 'Let's do the plan from the north *and* the south.'"

Bob's excitement appears to have been additionally fueled by al-Samarrai's assertions regarding the possibility of a military-led uprising that could occur simultaneously with an offensive from the north. However, al-Samarrai's talk of assassinating Saddam himself in Samarra gave him pause. Assassination of a foreign leader was clearly against U.S. law, and while the Bush administration had set out to do just that in the Gulf War bombing campaign, the targeting of the Iraqi leader had been legally, if flimsily, cloaked in euphemisms about aiming at "command and control centers." Involvement in a plan to gun down Saddam in his car would have no such excuse, certainly not for a CIA field officer acting on his own initiative.

According to a former colleague in the CIA, "Bob got scared. He tried to cover himself by reporting back about Samarra. That message got into a system [of communication] that reached the NSC." As we shall see, this report was to cause Bob a lot of trouble in the near future.

Meanwhile (according to statements by Chalabi, other senior INC officials, and Hoshyar Zibari of the KDP, which Bob denies), Bob set out to enlist Iran in his scheme to mount an offensive against Saddam's forces. This was very much forbidden territory, since standing orders strictly warned against any contact by CIA officers with the Iranians. Bob was not so reckless as to make direct contact with his opposite numbers in Iranian intelligence (though he did confer with the Iranian-backed Shiite group) but, according to Chalabi, he did the next best thing.

"Bob came to me and said, 'I have a message for the Iranians from the White House: "The United States would not object to Iran joining in the fight against Saddam Hussein provided it is committed to the territorial integrity of Iraq." ' " Bob denies making the statement and Chalabi recalls that he did not believe the message, which was hardly surprising, since it suggested a sudden abandonment of "dual containment," a cornerstone of the Clinton administration's foreign policy. "I said, 'I can't do that,' " reports the INC leader, "but Bob insisted."

As it so happened, "the Iranians" were close at hand. Like the CIA, Tehran's intelligence services kept a close eye on developments in Salahudin. Like the CIA team, they were strictly forbidden to talk to their American counterparts. In late February 1995, two such officers from the Pasdaran, a part of the Revolutionary Guards (an especially influential military and intelligence arm of the Islamic regime), were in the town. Chalabi went to them and said that while the CIA man could not talk directly to them, he had given Chalabi a message to pass on. He then related the contents of Bob's electrifying communication, adding that "I cannot vouch for the authenticity of the message."

Despite this caveat, Chalabi cooperated in a little bit of theater, in which by prior agreement the CIA officer made a public appearance in the lobby of the al-Khadra Hotel in Salahudin while the Iranians were visiting Chalabi as an implicit guarantee that the message from the White House was authentic. "Neither side was allowed to talk to the other, but they spoke in body language!" chortles an observer of the mime. "They were all standing there in the lobby; Bob eyed the Iranians, the Iranians eyed Bob. It must have gone on for three or four minutes."

This exercise in mute communication worked, at least for a while. The Iranians, in a state of high excitement, rushed to report the momentous news of the U.S. message to General Mohammed Jaafari, the presiding official on Kurdish affairs for Iranian intelligence.

By now, the welter of intrigue swirling in Salahudin and the area around it was becoming increasingly complicated. Various players appear to have convinced themselves that they were the ones manipulating events. Bob was convinced that he was the moving spirit in pushing Chalabi into action. "I told him he was wasting our money and time, but, more important, he was wasting a historical opportunity," Bob later claimed in a newspaper interview. "He knew I was right." Chalabi has not been loath to support this version of events, agreeing that "Bob kept pressing, 'When is this going to happen?'"

Others are not so sure, suggesting that, in fact, it was Bob who was being manipulated by Chalabi, who in turn was taking a lot of advice from Wafiq al-Samarrai. Hoshyar Zibari, the senior adviser to Massoud Barzani, states that "Wafiq, whom we had been paying for years before he came north—his information was good, by the way, or at least better than we got from others—was discussing a plan for some sort of coup with us, with the idea that we should make a military offensive as a cover. Chalabi sold the idea of the offensive to Bob in February 1995. Chalabi owed a lot of money that he had borrowed in late 1994 from businessmen in northern Iraq and he had to do something."

A former close associate of Chalabi's agrees: "Chalabi was using Bob as a tool rather than believing everything he said and promised."

All concerned—with the emphatic exception of the hyperactive CIA officer—are agreed that Bob now made a quantum jump in the audacity of his promises. Iraqi tank units, he announced, would be defecting to fight on the rebel side. According to Chalabi and Zibari, as an incentive to the various groups involved, particularly Barzani's KDP—militarily by far the strongest faction in Kurdistan—he promised that the military offensive would have American air support, a virtual guarantee of success.

Bob now claims that he never promised air cover. "He gets very angry if you mention it," says Chalabi. "But he certainly encouraged Barzani to believe that there would be U.S. air cover. When Barzani asked him if there would be air cover, I heard Bob say, 'Yes.'"

In one area at least, Bob did score a notable success. Energetically throwing his weight into the mediation effort, he and Chalabi persuaded the two warring Kurdish factions to declare a cease-fire. The hope was that they would now join forces against Saddam. But the most powerful of the Kurds, Massoud Barzani, remained highly dubious about the undertaking.

"Bob lied to everybody," says Hoshyar Zibari flatly. "He came to see us and said, 'I represent the President of the United States. I am here to execute his plan.' He promised air support." Barzani, definitely no fool, was the very last person in northern Iraq to take a CIA promise at face value. The memory of the American betrayal of his father in 1975 had not dimmed with time. (Curiously, one former CIA officer insists that Bob was probably the only agency officer in northern Iraq "who had a real understanding of what Barzani had been through and why he had good reason to mistrust the United States.")

At a meeting in late February, according to sources in the Iraqi opposition, Barzani made his misgivings clear. "There must have been a major change in U.S. policy," he remarked pointedly to Bob, "because every official who has ever come here has always told us not to provoke the Iraqis." He followed up this pertinent observation with some sharp questions.

"How will you defend Mosul and Kirkuk once we have taken them?"

"With our planes."

"How will the planes differentiate between Iraqi units that come over to us and the ones that are still loyal to Saddam?"

The Kurds were amused by Bob's quick reply: "We have given our secret allies in the garrison in Mosul a special paint to put on their tanks and trucks so that the planes will recognize them."

Barzani's suspicions were not allayed. He dispatched his close and trusted aide Zibari to London to check out what he had been told. Zibari called Washington, eliciting cries of horrified astonishment at the news of what was being promised coupled with fervent denials that the United States was contemplating military action of any kind in northern Iraq. Barzani drew the appropriate conclusions. Any offensive would go forward without KDP support—a decision he neglected to communicate to his allies.

It is an open question as to how much Bob's superiors really knew about his activities and when they knew it. Just possibly, as he insists, he kept Washington fully informed at all times of what was afoot. It is possible that they found out about the alleged promise of air support and the discussions with al-Samarrai about assassinating Saddam when the National Security Agency (NSA) intercepted a radio report from the Iranian intelligence officers in Salahudin to their superiors regarding their encounter with Bob. Zibari's queries would also have let the cat out of the bag.

In addition, Chalabi's rivals for the affections and support of the CIA were determined that there should be no risk of the INC scoring a success. The offensive was scheduled to begin on March 3, 1995. Two days before, General Adnan Nuri, leader of the Iraqi National Accord in northern Iraq, flew to Washington to drip poison in the ears of his agency handlers. The INC operation, he declared, was a devious plot masterminded by Wafiq al-Samarrai to draw the United States into another war with Saddam. He claimed that al-Samarrai himself had attempted to enlist him in the scheme, saying, "Come, and we make a plan to deceive the Americans."

Whether or not Nuri's malign spin was accepted, the highest levels of the U.S. government reacted with horror to the news from Kurdistan. A full-scale offensive by the opposition could well draw Saddam's forces north in a counterattack. The last thing the White House, the Pentagon, or the CIA wanted was to have to make good on the U.S. commitment to protect the Kurdish enclave in the north.

The offensive was timed to begin at midnight on March 3. That morning a cable arrived over the CIA communications system, dispatched by the President's national security adviser, Tony Lake, and addressed to the INC leaders. It fell to Bob to deliver the message. This must have been an unappealing task, since the message stated that "the United States would not support this operation militarily or in any other way."

Despite this depressing news, Chalabi did not give up. Addressing his commanders on the eve of battle, he declared that they were fighting for the liberation of Iraq and that they should press on regardless. The order was given to advance on all fronts.

The attack was to be on two fronts. In the east, Jalal Talabani's PUK Peshmerga would strike out for Kirkuk. A hundred miles to the west, a combined force of some ten thousand KDP Peshmerga and the INC's one thousand–strong militia were to advance toward Mosul. The front line opposite Mosul lay along the Great Zab River, a tributary of the Tigris, and was manned by KDP forces. When the small but eager INC body of troops reached the river, the KDP refused to let them pass.

Realizing that he had been abandoned for the second time in one day, Chalabi rushed to Barzani's gaudily ornate personal headquarters at Sara Rash, a few miles outside Salahudin, only to be greeted with the news that Barzani was out of town and on his way to Turkey. The anxious INC leader had to make do with the famously well-dressed Nachirvan, and despite pleading through the night, Chalabi failed to get the KDP to move.

Meanwhile, though Barzani had decided to sit out the offensive, Jalal Talabani and the PUK remained fully supportive. He may have agreed to press on in pursuit, despite the evaporation of American and KDP support, merely with the limited objective of pushing the Iraqis farther away from Arbil, the prize he had gained in the December fighting.

Initially, the offensive went well. Seven hundred Iraqi troops— "who probably hadn't been fed in two weeks," as one American humanitarian relief official observed—surrendered. The lightly armed groups of Peshmerga "occupied" a few square miles of countryside. There were no uprisings in any Iraqi military units elsewhere in the country. Just over two weeks after it began, the attack was over. On March 19, large numbers of Turkish troops crossed the border of Iraqi Kurdistan from the north. They were officially in pursuit of PKK guerrillas fighting the Ankara government from bases inside Iraq, though they may in reality have been responding to a request for help from Saddam. In any event, Talabani hurriedly pulled back his forces to protect Arbil, which lay in the line of the Turkish advance, and the front line reverted to its previous position. Even the prisoners had to be let go, as the INC could not afford to feed them. At least the commander of the Iraqi division in front of Kirkuk was sacked, in disgrace—a belated example of the "stick" approach in action—though he did not defect.

The whole sorry affair was a disaster for Ahmad Chalabi, the Iraqi National Congress, and anyone who hoped that Saddam *and* his regime could be displaced, with help from the outside, by an uprising. The CIA hierarchy had never been among that particular school and now they vented their irritation on Chalabi for his presumption in attempting such a bold stroke. INC stock at Langley, already in decline, now went into a deep slump. There were to be no more attempts at undermining Saddam from the periphery. From this point on, the agency devoted the bulk of its attention to fostering the long-anticipated coup, launched from within the Iraqi ruler's inner circle. Bob's dutiful report of the Samarra assassination idea, which had also reached the ears of the Iranians, gave the White House an excuse to punish him. He spent much of the following year under investigation by the FBI on charges of conspiring to murder a foreign leader. Ultimately, the Justice Department decided not to prosecute. The "military option," as CIA officers termed the increasingly popular coup plan, would of course almost inevitably result in the violent demise of Saddam Hussein.

As the spring of 1995 turned into summer and recriminations flew back and forth both within and between the CIA and Iraqi opposition groups, a far more vicious dispute was gathering force at the court of Saddam Hussein. "Cracks in the inner circle," long advertised by hopeful American spokesmen, were about to become a dramatic reality.

EIGHT

Deaths in the Family

The convoy of black Mercedeses had been driving fast through the night across the empty Iraqi desert for five hours when the white concrete arch marking the border with Jordan appeared in the headlights. It was the night of August 7, 1995, and Lieutenant General Hussein Kamel, long one of the most powerful men in the country, was fleeing into exile. With him he was bringing not only his younger brother, but his wife, Raghad, and sister-in-law Rina, two of Saddam's much loved daughters, not to mention their children and fifteen friends and relations from the Majid family. The world was about to learn of a momentous and unprecedented crack in Saddam's inner circle.

The fleet of cars roared toward the border post, the headlights briefly illuminating the life-size statue of Saddam Hussein that stands watch over the frontier, and briefly slowed down. The border officials took one look at the august group of travelers and respectfully waved them on their way. As the motorcade sped under the arch into Jordan, they left behind a regime consumed with the increasingly bitter hatreds and feuds of the ruling family. Even as the group raced down the narrow road to Amman, the unbridled rancor was exploding in gunfire and bloodshed.

Infighting within the family had been growing more intense since early in 1995, exacerbated at every stage by Uday. The previous year he had directed the strident media campaign against government officials that culminated in the assumption by Saddam himself of the post of prime minister. There had been no improvement in the general situation. Iraq's political and economic isolation continued. Bombs were going off in Baghdad. There were increasing signs of unrest among once fiercely loyal Sunni tribes, especially the powerful Dalaim, centered in the city of Ramadi, west of Baghdad on the upper Euphrates. After what may have been a conspiracy to assassinate Saddam sometime early in 1995, General Mohammed Mazlum al-Dalaimi was arrested. In May, the government handed his body, mutilated from torture, back to his relatives. Outraged, his fellow tribesmen rioted and attacked police stations. It took the dispatch of elite troops by Saddam and the death and wounding of several hundred Dalaimis before the trouble subsided. (A detailed and colorful report of a further uprising by Dalaimis in the army the following month was almost certainly disinformation, possibly part of a CIA strategy to create an ambience of "coups and rumors of coups.")

Uday had mercilessly criticized these security lapses in the press, criticisms that were by implication direct attacks on his Ibrahim uncles, Watban, the interior minister, and Sabawi, chief of the al-Amn al-Amm general security service, Saddam's half-brothers by his mother's second marriage. In May, Watban was dismissed.

But Uday was also directing his fire at the Majids, cousins on Saddam's father's side, traditionally rivals of the Ibrahims. Following attacks on Ali Hassan al-Majid, notorious as the hammer of the Kurds and defense minister since late 1991, this ferocious henchman of the ruler was dismissed from the defense ministry in the middle of July 1995. It is hard to say whether Saddam himself was actively encouraging these assaults on men so close to him, or whether Uday was running out of control. A week before Ali Hassan's dismissal, in what may have been a veiled reference to those relatives falling out of favor, Saddam criticized unspecified individuals who placed obstacles in his way "at a time when we were removing one arrow after another from between our ribs."

In any event, the ruler did not rein in his son, who now began to encroach on the military prerogatives of Hussein Kamel. As we have seen, Uday had been competing with the once all-powerful Kamel for some years in the business sphere. Now, two weeks after the downfall of Kamel's uncle Ali Hassan, Uday suddenly appeared to be moving to take charge of military transport by publicly overseeing the repair of military vehicles. On August 3, further highlighting his newfound interest in military affairs, Uday attended an air show. This was a fight over money as well as authority. At this precise time, Hussein Kamel was trying to set up a military contract with an Eastern European country. Uday wanted to muscle in.

Kamel later said that he and his brother, now a lieutenant colonel in the Amn al-Khass presidential security service, decided to flee the country at the end of July. He said he talked to Raghad, with whom he now had three children, and Rina (Saddam Kamel's wife). "Ten days before we decided to travel, I explained all the details without any hesitation. Perhaps at the beginning we thought that they might tell their family. But I did not care about that and said: 'You either get ready to travel with me or I will travel alone.' They did not mind at all and came along with me to Amman."

Kamel had certainly laid careful plans for his escape. For some time before he set off on that fateful trip across the desert he had been sending his accountant around the headquarters of the various government organizations he controlled with requisition slips for whatever hard currency they happened to have in their safes. None could refuse the emissary of the apparently all-powerful Kamel, and as much as several million dollars were collected in this fashion.

"I am a known person," Kamel said later with the arrogance that never deserted him. "No soldier could stop me on the road." Even so, he waited until night fell on August 7 before setting off, either to escape notice in the darkness or simply to avoid the searing daytime heat of midsummer in the western desert. He must have known that that date held deep significance for the Saddam regime, since it was the eve of the anniversary of the victory over Iran in 1988, a victory in large part gained by the elite Republican Guards divisions that he himself had founded. The departing general may also have known that many of the Iraqi ruling clique would be celebrating at a party at a country house outside Baghdad.

It was a party that Kamel was fortunate to miss. What happened was in many respects a rerun of the night on the "Mother of Pigs" island in the Tigris seven years before, when Uday had murdered Kamel Hannah Jajo, his father's aide and pimp, in a fit of drunken rage. Among the senior members of the regime present at this festivity was the ruler's half-brother Watban Ibrahim. Also present was Uday's cousin and boon companion Luai, the young man who had once had his arm broken by Saddam for kidnapping and beating up his schoolteacher.

There are various reports of what caused the party to turn violent. According to one account, Watban's son Ahmad quarreled with Luai and slapped his face. Luai called Uday, who raced out to the party with his submachine gun, arriving at about three-thirty in the morning. Other reports assert that Uday's angry arrival was prompted by the news that Watban had been speaking ill of him. In any event, his reaction was extreme. Bursting in on the festivities, he sprayed the room with gunfire. The hail of bullets hit Uncle Watban, severely wounding him in the leg, as well as killing six young women, gypsy dancers and singers considered essential by Tikritis for any social occasion. (Later that morning, Ahmad retaliated with the feuding vigor of a true Tikriti by firing a rocket-propelled grenade at Luai's father's house.)

Even as Uday's victims were being transported to the hospital or morgue, Hussein Kamel and party were checking into the al-Amra Hotel in the center of Amman, the Jordanian capital.

Jordanian government officials later claimed that his defection came as a complete surprise. "Only Hussein Kamel and his brother knew what they were going to do," says Abdul Karim al-Kabariti, Jordanian foreign minister at the time. "We knew he had crossed the border, but it was not unusual for an Iraqi official to enter Jordan without giving us any information about what he was doing here." Al-Kabariti, a svelte, intelligent, former banker, had met Hussein Kamel a few years before and had disliked him immediately. "He tried to impress on me how anybody could build a nuclear weapon. It was all a matter of will and funds and natural materials." The foreign minister only learned the astonishing reason for the Iraqi general's return to Jordan when King Hussein called him to the palace, where he found other senior ministers already assembled. The king told them that Hussein Kamel had just telephoned the official in charge of the royal court from the al-Amra to say that he and his family planned to seek political asylum in Jordan.

The king was not being entirely forthcoming with his ministers. At some point before Kamel set off, he had contacted the Jordanian monarch and intimated what was afoot. The king in turn relayed a cautious message to Washington that, in the words of a former CIA official who was privy to the message, "something big was going to happen" in Iraq.

Keeping the secret of Kamel's plans and welcoming him when he arrived was a critical decision for King Hussein. He was Iraq's closest friend in the Arab world. He had allied his country with Saddam during his war with Iran. Iraq was Jordan's biggest market. He had been a friendly neutral during the Gulf War. The long road between Amman and Baghdad, down which Hussein Kamel had just driven, was Iraq's only outlet to the world. The majority of Jordanians are of Palestinian origin. They sympathized with Saddam Hussein's assault on the established order in the Arab world in 1991. They applauded Iraqi missiles fired at Israel. For months, Saddam's picture decorated every shop and taxi in Amman.

Jordan had paid heavily for its friendship with Baghdad. Most of the 350,000 Palestinians in Kuwait, expelled after the Gulf War because Kuwaitis saw them as pro-Iraqi, had moved to Jordan. Saudi Arabia, Kuwait, and the Gulf states, which had previously subsidized Jordan, became hostile. The king had told Jordanian editors in 1993 that "Saddam has broken our backs." The king had initiated his return to America's good graces by signing a peace treaty with Israel in 1994. Now he decided to make a final break with his old ally in Baghdad. He told his ministers: "Things can't be tolerated with Saddam anymore."

Two days after the king had granted Kamel's request for political asylum, Uday and Ali Hassan al-Majid arrived in Amman demanding to see King Hussein. Hussein Kamel warned the Jordanians about the possibly murderous intentions of his relatives, particularly his uncle Ali. He said: "Don't let his majesty shake hands with this man. He might have something in his hand that might kill him." King Hussein did not think he had much choice but to meet the two Iraqi emissaries.

They asked for the extradition of the defectors, but must have known this was a lost cause. Their main interests were in retrieving Kamel's bank card—evidently his fund-raising initiative prior to leaving had been noticed—as well as in seeing Raghad and Rina, whom they claimed had been brought to Jordan against their will. King Hussein turned them down. He said: "My daughters spend time with them. They want to stay." He promised to look after them.

Uday and Ali Hassan went back to Baghdad empty-handed. They had seen how King Hussein was using the opportunity offered by the Kamel brothers' defection to turn against Saddam. The king praised Hussein Kamel in an interview and said it was "the right time for change" in Iraq's leadership, adding that "if a change occurs it will only be a change for the better." The new direction in Jordan's diplomatic allegiance was underlined by the fact that the king chose to announce his new stance in an interview with *Yediot Aharanot*, an Israeli newspaper. President Clinton phoned him to promise to defend Jordan against Iraqi retaliation.

The defection of Hussein Kamel had caused an international sensation. Commentators around the world eagerly interpreted his dramatic departure as a sign that Saddam's regime was a "sinking ship." When, after four days in seclusion, this erstwhile pillar of the mysterious and frightening regime in Baghdad made his first public appearance at a press conference in the garden of one of the king's palaces, it indeed appeared that Saddam had gained a formidable opponent. Dressed in a double-breasted gray pin-striped suit, he gave a résumé of his career and declared: "We are working to topple the regime." He made clear that he was speaking of a coup and not a popular insurrection by appealing to "the entire army, Republican Guards, and Special Guard officers." He was factual and well informed, accusing the regime of leading Iraq into "complete isola-tion." He did not attack Saddam and his family personally "because of kinship." He said he would not "take responsibility for unveiling any secrets." He was not unimpressive, but he seemed like a man who had been in power too long to make a good revolutionary.

Despite his sudden metamorphosis into a critic of Saddam, Kamel was hardly likely ever to succeed as an opposition leader. A leading member of the Iraqi regime until just days before, he had shared in its crimes. Jaundiced though Barzani was on the subject of Hussein Kamel, he had a point when he said: "Between him and the Kurds there are deep wounds. When a Kurdish delegation came to Baghdad [in 1991], he was the harshest in attacking them and calling them agents. . . . How can the Shiites of Iraq deal with him when he attacked the tomb of Imam Hussein bin Ali?"

In calling for the overthrow of Saddam Hussein, Kamel appealed to the security services and the army. But the place for him to lead such a coup was Baghdad. The very fact that he and his brother had fled to Amman showed that they did not really believe in a military uprising.

Nevertheless, even if he was not the man to lead the overthrow of his father-in-law and former master, Kamel represented a tremendous intelligence catch and Jordanian, Arab, and Western intelligence organizations were eager to speak to him. "They wanted information and he provided it," says Kabariti. "But it wasn't up to expectations. The Kuwaitis thought he would tell them about hostages [who disappeared during the Iraqi occupation of Kuwait]. The Saudis thought he would tell them about Iraqi plans. The Americans thought he would brief them about Iraqi weapons of mass destruction. One could consider that either he didn't have the information—or wanted something in return."

That was not entirely true. Certainly, the arrogant general's interview with the CIA did not go well. He felt insulted that the officers sent to interview him were not of high rank, nor did they speak Arabic. Instead, the CIA team had brought along an interpreter of Egyptian origin, who found Kamel's Tikriti accent difficult to understand. For their part, the agency officers concluded that he was "just an idiot," as one of their colleagues later recalled. "His plan was that he would return to Baghdad behind the U.S. Army and Air Force. End of subject."

Another interviewer had better luck. Rolf Ekeus had first met Hussein Kamel in June 1991, when the Iraqi was at the height of his power. On that occasion, Kamel had rudely interrupted a meeting between the Unscom chairman and some of Iraq's more suave diplomats. In the years since then, Kamel had been in charge of the elaborate effort to obstruct Unscom and conceal as much as possible

of Iraq's weapons of mass destruction programs. Now the two former antagonists were meeting under very different circumstances.

Their first meeting in Amman, almost two weeks after Kamel's
defection, began with a startling revelation. On entering the room,
the Iraqi looked carefully at the faces of the Unscom personnel sitting on Ekeus's side of the table. Finally his gaze settled on the
Unscom chairman's Arabic interpreter. "Are you a Syrian?" he
asked. The man admitted that he was. Kamel asked: "Is your name
Tanous?" When the visibly nervous translator said this was correct,
Kamel replied, "Get the fuck out of here. You have been working for
me. I refuse to be debriefed by one of my own agents." As he
explained, the man had been infiltrated into Unscom employment
and had long provided much useful intelligence. Jordanian security
officials monitoring the meeting were much amused.

Once the Iraqi mole had been removed, Kamel was anxious to be
accommodating. "We have been enemies before," he said to Ekeus.
"Now we meet as friends." Before Ekeus could turn the discussion
to weapons, Kamel wanted to complain about his brother-in-law.
Uday spent his life in bars, picking fights, drinking, chasing women.
He, on the other hand, complained the general, worked long hours.
He was a teetotaler and a family man. The whole Saddam family, he
explained, squeezed by sanctions and violated by Unscom inspections, was "full of hatred. They are boiling with hatred."

Having gotten that off his chest, Kamel turned to Ekeus's area of
interest. He had been surprised, he said, at the effectiveness of
Unscom, something none of the Iraqi leadership had expected when
the inspectors first arrived. Ekeus, for his part, was interested primarily in the methods employed by Kamel's men in their task of
concealing weapons, materials, and documents from the intrusive
searches of the Unscom team.

"One of my first questions was 'How did you do it?'" the Swedish
diplomat recalled later. "He was eventually quite forthcoming, and
so were some of the officers who had come with him, even though it
was hard for them to change" from the habits of secrecy about such
matters ingrained in all Iraqi security officials.

Ekeus was being discreet. One in particular among those al-Majid
officers in the Kamel party supplied the most important information.

Major Izz al-Din al-Majid was the Special Republican Guard officer in whose Abu Ghraib villa garden the priceless parts and tools from Project 1728 had been buried in July 1991. As a member of the ruling family and an officer in the elite security unit, Izz al-Din was one of the select few chosen to move and hide, out of Unscom's reach, the forbidden weapons and materials. As such, he was able to furnish the interviewers with crucial insights into the way the concealment system functioned and who was involved.

Kamel's defection had in any case yielded a rich dividend for Ekeus even before the two men met in Amman. Three weeks prior to the flight of the Kamel brothers, Saddam Hussein had made a defiant speech in Baghdad in which he threatened that Iraq would cease all cooperation with Unscom if there were no progress in the Security Council toward the lifting of sanctions. When Ekeus met Tariq Aziz in Baghdad just three days before the Mercedes convoy sped across the border, Aziz repeated the threat, adding that the deadline for the Security Council to change its ways was the end of August. The deep-voiced deputy prime minister also added that, while Iraq had indeed done research on biological warfare agents, the scientists had never succeeded in producing them in a form suitable for use in a weapon.

This was too much for Ekeus. "Of course you have," he interjected.

Aziz took a deep puff on his cigar, his normal reaction when confronted, then fell back on a familiar defense. "Iraq is not like Sweden, where you make a plan and implement it. We are incompetent."

On August 13, the day after Hussein Kamel's press conference, the Iraqi government performed an abrupt and dramatic about-face. Fearful that the traitor would earn some reward for himself by giving away their darkest secrets, the government resolved to beat him to the punch.

Ekeus, who by this time had returned to New York, got an urgent message from General Amer Rashid, the brilliant British-trained engineer who had been acting as Kamel's deputy in dealing with Unscom, asking Ekeus to come back to Baghdad as soon as possible. Furthermore, wrote Rashid, "the government had ascertained that General Hussein Kamel had been responsible for hiding important information on Iraq's prohibited programs from the commission and IAEA by ordering the Iraqi technical personnel not to disclose such information and also not to inform Mr. Tariq Aziz or General Amer of these instructions."

Ekeus returned to Baghdad, where he encountered Aziz, Rashid, and other senior officials all suddenly exuding goodwill and promises of cooperation. Everything, they explained, had been the fault of Hussein Kamel. The rest of the Iraqi government had been quite ignorant of his nefarious activities in concealing the forbidden weapons programs. Henceforth Iraq would pursue a policy of full cooperation with Unscom and "good-neighborliness" with other countries. In addition, it was now admitted that Iraq had not only succeeded in manufacturing biological weapons but had actually loaded them into 166 bombs and 25 al-Hussein missile warheads.

Nor was that all. As he was about to leave Baghdad, Ekeus complained that so far he had not seen a single document to back up all this interesting new information. Within less than an hour, Ekeus got a call from Rashid suggesting that on his way to the airport (closed under sanctions to all but UN flights) he and his team should stop by a farm belonging to Hussein Kamel, in a place called Haidar, where he would find "items of great interest." That was putting it mildly. In a locked chicken shed, Ekeus found piles of metal and wooden boxes packed with over half a million documents as well as microfiches, computer disks, and photographs. Almost all of this treasure trove carried an abundance of detail about the secret weapons programs, particularly the nuclear weapons effort. The Unscom group later concluded, after carefully analyzing pictures of the farm taken on preceding days from a high-flying U-2 spy plane, that the files they had discovered had been carefully purged of the most sensitive material in the twelve days since Kamel's flight.

Thanks to the debriefings in Jordan and the "chicken farm" documents, Ekeus and his team discovered how well the Iraqis had fooled them over the previous four years. Not only was Wafiq al-Samarrai's information on VX and biological weapons confirmed and amplified, but they had also learned for the first time of Project 1728 and a secret missile test that had taken place in 1993. They had also learned of the invisible organization dedicated to outwitting them that they came to call the "concealment mechanism."

Meanwhile, Ekeus reported on his arrival in Amman that there was "political panic" in Baghdad. Uday's Fedayeen militia were deployed on the road leading to Jordan. Hussein Kamel himself said that since he had gone to Jordan "there is not a single street in Baghdad without a Republican Guard position to search people." Ordinary Iraqis were electrified. They had never seen such a split in the ruling family. "People held parties throughout Baghdad," says one of those who celebrated. "They believed the regime was wobbling."

The most dangerous moment for Saddam was immediately after the defection. The defectors denied that they had organized a conspiracy while still in Iraq, but he could not be sure. He did not know how much support Hussein Kamel enjoyed in presidential security—the Special Guard and the Republican Guard. After their flight, Hussein Kamel said he expected arrests and executions. But the regime must have had some confidence that he could not orchestrate a coup from Amman, since it did not cut the telephone link to Jordan.

Meanwhile, Kamel was rapidly removed from his many important positions. On August 10, it was announced that he had been fired as industry minister and director of military industrialization. A week later, the Baath Party expelled him.

More menacing for the exiles was the strength of the denunciations from Saddam Hussein and the al-Majid clan. In a wordy and elliptical speech on August 11, the Iraqi leader compared Hussein Kamel, successively, to Cain, who murdered Abel; to Judas, who betrayed Christ; to Croesus, the infamously avaricious king of ancient times; and to Abu Lahab, who had opposed his nephew, the Prophet Mohammed. Saddam said his son-in-law would be "stoned by history" and would do better "to die than live in humiliation." He accused him of stealing several million dollars through front companies. He predicted, not wholly wrongly, that Hussein Kamel would be at the mercy of his new foreign masters and would have to obey them "without any argument or right of veto."

Even more damning was a statement from the al-Majid clan, signed by Hussein Kamel's uncle Ali Hassan al-Majid, denouncing Kamel for treason against Saddam. It also carried a very direct threat to his life, underlining, at the same time, how seriously Iraqis still took tribal law. "Although the traitor Hussein Kamel belongs to an Iraqi family," wrote the al-Majids, "this small family within Iraq denounces his cowardly act." They called for his punishment. His relatives formally announced that they would not seek vengeance against anybody who killed him, saying flatly: "His family has unanimously decided to permit with impunity the spilling of blood."

The Iraqi press denounced the traitor for taking advantage of the "tolerance" of Saddam Hussein and for stealing money. Most of the abuse was crude, but the media did publish one telling document, aimed at showing up Hussein Kamel as a poorly educated sycophant. This was a letter he had sent Saddam on October 13, 1994, when Iraqi troops were withdrawing from the border with Kuwait after a mini-crisis with the United States. It is in ungrammatical Arabic and contains several spelling mistakes. The note reads: "Dear Sir, It is not important that sanctions be lifted. What is important is to see the world mentioning your name everyday. Our hope is being materialized. May God be with your excellency and our souls are nothing before your excellency."

Meanwhile, Watban was in the hospital, where Cuban doctors were fighting to save his leg. It was a measure of the rigor with which the government still controlled Iraq that an Iraqi specialist who was also tending Watban had to return every night to prison. He was serving a six-month sentence for illegally erecting a television satellite dish to watch foreign broadcasts.

Uday's murderous assault on his uncle and the gypsy dancers went unmentioned in the Iraqi media, but for the first time, the prince was on the receiving end of official and public abuse. Mohammed Said al-Sahhaf, the Iraqi foreign minister, said he was "unfit to govern." Barzan, in Geneva, coupled Uday with Hussein Kamel, saying: "Problems come from people who do not appreciate what their true size is. The notion of inheriting power is not acceptable in Iraq. People do not accept Uday or Hussein Kamel. Neither of them has the legitimacy to govern."

Even Saddam was distancing himself, at least publicly, from Uday and clamping down on his empire. The Iraqi Football Association reelected Uday as its head by 155 to 0, but officials said he would be confining himself to sports. The Iraqi leader declared there was no room in Iraq "for a state within a state." Iraqi security raided Uday's Olympic headquarters and freed three people from his private jail.

One street rumor, possibly inspired by the government itself, told how Saddam had gone to his son's immense private garage. Shocked to discover that his son owned sixty luxury cars, he was alleged to have ordered his guards to sprinkle petrol over the vehicles and set them alight.

Saddam adopted an aggressive strategy to show that he was still in control despite the shooting of Watban by Uday and the defection of Hussein Kamel. The government announced that a referendum would be held on October 15 in which 8 million Iraqis would vote on the question: "Do you agree that Saddam Hussein should be president of Iraq?" While the result was not in doubt, the campaign focused attention on the Iraqi leader, still only fifty-eight and in good health, though he dyed his mustache and suffered from back pains. Hundreds of foreign journalists were invited to watch the voting, more than had been allowed into the country at any time since 1991. Even if their coverage was unsympathetic, they could see that the government had a tight hold on power everywhere in Iraq except Kurdistan.

The referendum campaign, organized with relentless efficiency by the Baath Party organization, also saw a further exaltation of Saddam's personality cult. "O lofty mountain! O glory of Iraq!" wrote Ali Hassan al-Majid. "By God we have always found you in the most difficult conditions a roaring lion and courageous horseman, one of the few true men." The deification of Saddam was evident all over Iraq. At one polling station in the Arafa district of the oil city of Kirkuk, there were thirteen pictures of the Iraqi leader on the walls. He was portrayed in different guises, such as an Arab sheikh, a baggy-trousered Kurd, and a white-suited businessman. A fourteenth picture of the leader was stuck to the ballot box. In a local primary school, there was a special board on which children had stuck love letters and birthday cards to the Iraqi leader. It faced a large mural depicting an Iraqi soldier in the act of repeating the words of his leader: "Victory is sweet."

Saddam won 99.96 percent of the vote. It was the first such vote since 1921, when the British had organized an equally spurious poll showing that 96 percent of Iraqis wanted Faisal I as their king. There were no alternative candidates in either 1921 or 1995. Saddam had a further objective. By curbing Uday and holding a referendum under the auspices of the party, he was saying, as one diplomat in Baghdad put it, that "in future, government will be in the hands of bodies like the Revolution Command Council and the Baath Party and not the inner family."

The inner family did not lose power. Qusay rapidly assumed command of the important offices vacated by Hussein Kamel. The essential levers of power were still the carrot and a very brutal stick. But Saddam had always shown an uncanny gift for balancing the administration of Iraq between loyal subordinates chosen for their unquestioning loyalty—usually family members—and highly skilled experts. Often he dispensed mercy as a reward, all the more welcome to the recipients because it was unexpected. For example, when Kamel left, his highly intelligent and capable deputy, General Amer Rashid, must have felt extremely nervous. His proximity to the departed traitor would certainly have laid him open to suspicion and the tender mercies of Qusay's interrogators. But, instead, Saddam promoted him to the post of oil minister as well as giving him Kamel's old job of dealing with Unscom. In addition, Rashid was entrusted with the task of trying to track down the millions of dollars Saddam believed Hussein Kamel had stolen.

Hussein Kamel could not have predicted one serious consequence of his flight from Iraq. His arrival in Amman gave Jordan an excellent opportunity to switch decisively into the American camp. The political isolation of Iraq was complete. In future, plots against the regime would come out of Amman and not, as hitherto, Iraqi Kurdistan.

In Amman, Hussein Kamel was not responding well to the pressures of exile. The king had lent him a house that had once belonged to a former wife who had died in a helicopter crash. Kamel lived there with Raghad, with whom he was getting on badly, and their three children, along with his brother, Rina, and their two children. He only once went out into the city—for a medical checkup in the hospital. He was bored and lonely. Alia, the king's eldest daughter, owned the neighboring palace in the royal compound. She found the Kamels always walking through her garden to borrow videos. The exiled couple would then hang around in her house for hours.

To escape their company, she finally fled to one of her other houses in Palm Beach, Florida.

The opposition, mainly based in London and northern Iraq, spurned Hussein Kamel. His appeal to the elite units of the Iraqi army to support him had not produced a single mutiny. After a few months, foreign intelligence services lost interest in debriefing him. The Jordanians, who had now settled on the exiled opposition group Iraqi National Accord as their chosen instrument to bring down Saddam, were no longer enthusiastic about the Kamels' presence in Amman. King Hussein pointedly invited Raghad and Rina to dinner, but not their husbands.

One of the few people who came to call was Rolf Ekeus, still winkling out information on the weapons programs. He found the sad decline in Kamel's fortunes mirrored in his surroundings. During their early meetings, the house had been a hive of activity. The phones never stopped, fax machines spat out endless messages, aides and emissaries bustled in and out. Now Kamel sat alone. A single broken air conditioner sent out a continuous rasping noise. There was a layer of dust everywhere and the phone never rang.

Hussein Kamel did use the telephone to talk to General Wafiq al-Samarrai who, following the debacle of the INC offensive in March 1995, had hurriedly made his way from Kurdistan to Damascus, where he enjoyed the protection and sponsorship of the Syrians. By October, Kamel was considering a move to Syria himself. Here he faced a problem. When he asked King Hussein for permission to go, he was told that he and his brother were free to travel to Syria, but without their wives and children. The king had promised Uday in August that "Saddam's daughters are my daughters." The Jordanians had already aroused the dangerous ire of Saddam Hussein by accepting Kamel and swinging into the American orbit. They did not want to provoke Saddam any further by sending the Iraqi ruler's own daughters into the custody of his most hated rival, Syrian president Hafez al-Assad. The king said: "The girls will have to stay here."

Relations between the Iraqi exiles and their Jordanian hosts became frostier by the month. On January 4, 1996, Kabariti gave an interview to the Amman newspaper *Dustur* in which he said that Hussein Kamel was "most welcome when he came. When he wishes to leave, we will treat him the same way." The general made a brief effort to establish himself as the leader of an opposition group, to be called the "Higher Council for the Salvation of Iraq." Its program was intended to appeal to the Sunni establishment in Iraq. It opposed Saddam Hussein, but renounced foreign aid to get rid of him. It pledged that there would be no witch-hunt after he was overthrown and promised elections, but not federalism. Kurds would get their natural rights within a unified Iraq. Nobody showed the slightest interest in the higher council or its program.

Abdul Karim Kabariti, who became prime minister of Jordan at the beginning of February, says, kindly, that Hussein Kamel's problem was not stupidity, but that "he could not do without power." He was used to giving orders and seeing them carried out. Even his first speech appealing for the support of the Iraqi army sounds like a commanding officer giving an order rather than a politician looking for support. Others, who came to know Kamel at least as well as Kabariti, were less impressed by his intelligence and abilities. "He had a reputation as an excellent manager, based on his work with the Republican Guards and other things," said Rolf Ekeus. "But the test of a good manager is his ability to operate with finite resources, to decide on options. Kamel operated simply by means of infinite resources and infinite ruthlessness. Otherwise he was an extremely stupid man."

One incident in particular brought home to the general his diminished status. He had begun to criticize King Hussein's plans to establish a pan-Arab, anti-Iraq front. He praised some minor reforms in Baghdad. Nayef Tawarah, the editor of the Amman newspaper *Bilat,* told Kamel of his intention of publishing some of his comments critical of King Hussein. The Iraqi wanted to stop him. Reverting to a mode of behavior customary in Baghdad, he threatened to kill Tawarah. "I will cut you up piece by piece," he told the journalist, who was taping the conversation.

Tawarah, a friend of Kabariti's, announced that he was going to sue. The Jordanian government told Kamel he would have to stand trial. Kamel riposted to Kabariti that the journalist's action was "inconceivable." The Jordanian minister piously observed that "We are all living under the law in Jordan." Kabariti, with a certain glee, recalls: "He really couldn't believe it. His face went pale, in fact, yellow. He clutched a pillow to his stomach. He kept repeating, 'This is unbelievable, unbelievable, unbelievable.'"

Saddam had predicted immediately after Hussein Kamel's flight that his newfound friends would suck him dry of information "until he is burnt out and then throw him into the road." Now he set to work with chilling skill to seduce Kamel into returning. He sent assurances through Kamel's father and to Raghad, through her mother, Sajida, that the defectors could return in safety to Baghdad. On at least one occasion, he called Kamel directly with assurances that the prodigal son-in-law need fear no repercussions. "Do you think I could harm the father of my grandchildren?" asked Saddam with dramatic sincerity. Unbelievably, Kamel began to take him seriously.

There were prolonged negotiations. Baghdad was obsessed by the belief that Kamel had built up an enormous fortune abroad through commissions from when he had been in charge of Iraqi military procurement. Barzan gave a precise figure. He said: "Between 1985 and 1995, he controlled seventy-three percent of Iraq's funds." When Uday and Ali Hassan al-Majid failed to retrieve his bank card when they came to Amman after his defection, they went to the trouble of canceling it. These negotiations about money may have diverted Hussein Kamel from considering the specific threats to kill him made six months earlier.

On February 19, Hussein Kamel sent a formal letter to his father-in-law asking about his return. He told reporters that "the initial response was positive." Not all his family agreed with his decision. For once, Saddam Kamel, who had made little impact on anybody in his six months out of Iraq, protested vigorously. "You donkey," he reportedly shouted at his brother. "You want us to go back to our deaths."

In reply, Hussein Kamel pulled out his pistol and said, "You will come back." Izz al-Din al-Majid called a Jordanian intelligence official from Turkey, where he was visiting. "What about my kids? They will be killed," he said plaintively. The Jordanians told him to send them a fax saying he did not want his children to go back to Baghdad. Nothing arrived. When they spoke with him again, he said resignedly: "Let them go back. Leave it to God."

Kamel's decision to return, knowing what he did about his father-in-law's attitude to anyone who betrayed him, has long mystified even those with only a passing interest in Iraq and its malignant ruling family. A friend who spoke to him near the end of his brief exile suggests that Kamel was "driven mad" by King Hussein and his advisers. In particular, the decision to let the journalist's lawsuit go forward convinced the distracted general that the king intended to jail him. "Better to be killed by my relatives than to rot in a Jordanian jail," he said.

The Jordanians did have some last-minute qualms. On February 20, King Hussein called Kabariti to say that Hussein Kamel had been to the Iraqi ambassador's residence and was now at the Iraqi embassy—a kitsch Babylonian building in central Amman. "Shall we let him leave?" asked the king. "Let him go," answered the politician. "It will be a great relief." By now, Hussein Kamel and his relatives were loading their belongings into the same Mercedes sedans that had brought them to Amman seven months before.

Hussein Kamel himself must have had misgivings about the wisdom of his actions during the four-hour drive through the stony desert of eastern Jordan. It is a depressing road, narrow and dangerous because of the large trucks traveling between Amman and Baghdad. Every half hour he said to the driver: "Stop. I want to take a piss." The driver reported later that he would stop and Hussein Kamel would "get out and pace up and down as if he were making up his mind, but he would not piss."

It is not clear when Hussein Kamel and his brother realized that they were going to die. The Iraqi government said that on his arrival at the border at Trebeil "the leadership took a decision to accept his appeal . . . to return as an ordinary citizen." More ominously, Uday was waiting for him. He did not try to arrest the Kamel brothers, but he took his sisters Raghad and Rina, with their children, into his motorcade.

From the other side of the border, a Jordanian security official was watching closely, with an open line to the king's palace in Amman. The moment he saw Kamel being separated from his family he reported back: "*Khallas*"—"He's finished."

Left to their own devices for the time being, the brothers drove to a house they owned in Tikrit. When they arrived, they found their al-Majid relatives angry and threatening. Not daring to spend the night there, they drove south along the Tigris to Baghdad, to the house of their sister, who, along with her children, had returned with them despite the misgivings of her husband, Izz al-Din. Here they were joined by their father, and Hakim, their youngest brother.

Then came a summons to the presidential palace. Uday's friend Abbas Jenabi, the editor of *Babel*, later recounted that an angry Saddam demanded that the brothers sign documents immediately divorcing his daughters. When they refused, one of the family threatened to shoot them on the spot, but Saddam intervened, saying they should have two days to reconsider. On February 23, Uday's television station announced that Raghad and Rina had divorced their husbands. The language of the announcement showed that Hussein and Saddam could expect little mercy. It claimed, as Uday and Ali Hassan al-Majid had said to King Hussein seven months earlier, that Saddam's two daughters had been brought to Amman against their will. It said they had told the king that "they had been deceived and misled by two failed traitors." It concluded by announcing that Raghad and Rina were "refusing to stay married to men who betrayed the homeland, the trust, and the lofty values of their noble family and kinfolk."

By now Hussein Kamel must have known there was no escape. Along with his father and two brothers, he waited in his sister's villa in Baghdad. A squad of forty men from the presidential guard, of which Hussein and Saddam had once both been members, surrounded the house. Given that Saddam had recruited his bodyguards from close relatives, it would not have been difficult to ensure that they were all members of the al-Majid family. They were led by Ali Hassan al-Majid himself.

In bizarre deference to the proprieties of tribal feuding, the assault party sent ahead a Honda filled with automatic weapons and ammunition for the Kamel family to defend themselves with. It would be a fair fight. Uday and Qusay watched the proceedings from a car parked nearby.

When the assault began, Hussein, Saddam, and Hakim Kamel fought back fiercely from the house. The battle went on for thirteen hours, during which time the Kamels succeeded in killing two of the attackers. When they ran out of ammunition, Hussein Kamel, who had already been wounded, came staggering out of the house and shouted: "Kill me, but not them." He was shot dead. Saddam Kamel was hit and killed by a rocket grenade as he fired from the balcony of the villa. His father, Kamel Hassan al-Majid, his sister, and her children died inside the house. When the fighting was over, Ali Hassan al-Majid stood over his nephew and gave him one last shot in the head, saying: "This is what happens to all those who deal with the midget" (a reference to the diminutive King Hussein). According to one story current at the time in Baghdad, the attackers then put meat hooks in the eyes of the dead brothers and dragged them away.

Babel TV, Uday's station, got the scoop on what had happened. It quoted a spokesman from the Interior Ministry as saying that "a number of young people from the Majid family" had killed the three Kamel brothers. The official line was that the state might have pardoned them, but not their own tribe. Later the al-Majid clan put out a statement saying: "We have cut off the treacherous branch from our noble family tree. Your amnesty does not obliterate the right of our family to impose the necessary punishment." Saddam himself said later of the al-Majid family's actions, "Had they asked me, I would have prevented them, but it was good that they did not." The next day, Uday, wearing tribal robes, walked in the state funeral procession of the al-Majid clansmen killed in the battle at the Kamel house.

Raghad and Rina, once Saddam's favorite children, never forgave him for the killing of their husbands. They assumed he had orchestrated the attack by the al-Majid clan. They continued to live with their five children in a family house in Tikrit, never going out, always wearing black, and refusing to see any member of their family apart from their mother.

Saddam had survived what might at first have been a crippling blow to his regime. Hussein Kamel's defection had signaled a definitive crack in the inner circle that ruled Iraq, and yet the apostate son-in-law had been neutralized and ultimately eliminated with comparative ease.

Yet the dramatic episode of Hussein Kamel's flight had shifted the axis of the Western intelligence agencies working to bring down Saddam. Now they were working from Jordan, and over the next few months they were to make their most determined bid so far to destroy him.

NINE

"Bring Me the Head of Saddam Hussein"

T he amateur cameraman focused on the dark-suited man with a scraggly mustache sitting behind the paper-strewn desk. It was midwinter, and the Zagros Mountains of northern Iraq, visible through the window behind the desk, were topped with snow. Just outside, traffic thronged the busy street in downtown Sulaimaniya, the capital of eastern Kurdistan. Staring fixedly into the lens, his rodentlike face showing signs of extreme nervousness, Abu Amneh al-Khadami began to recount his career as a terrorist bomber on the CIA payroll.

No one had ever claimed responsibility for the bomb blasts that echoed around Baghdad in 1994 and 1995. One explosion had gone off in a cinema, another in a mosque. A car bomb outside the offices of *al-Jumhuriya*, the Baath Party newspaper, had wounded a large number of passersby and killed a child. Altogether, the bombs had killed as many as a hundred civilians. As we have seen, Uday put the blasts to political use, publicizing them as a means of undermining his uncle, interior minister Watban Ibrahim. Now, on January 25,

1996, Amneh had brought the video camera to his office to record the history of his role in the lethal blasts.

For the next hour and a half, he talked steadily, pausing only to light cigarettes, occasionally holding up operational orders from his paymaster to illustrate the story. This was not a confession by a repentant murderer, but a rambling complaint that his work for the cause had been impeded by lack of explosives and money.

The bombings, claimed Amneh, had been planned and executed on the orders of "Adnan." He was referring to Adnan Nuri, the former general in Saddam's army who had been recruited by the CIA in 1992 to work directly for the agency. Since that time, Nuri had risen to command the operations of the opposition Iraqi National Accord in Kurdistan. His mission, as mandated by the CIA, was to work on preparations for a coup inside the Iraqi military that would, finally, eliminate Saddam.

Nuri had recruited Amneh from a jail in Salahudin, where he had been incarcerated by Massoud Barzani's KDP for attempting to kill an official of the INC. He claimed his release was due to direct intervention by the CIA, quoting Nuri's boast that he "made the American in Washington telephone Massoud Barzani to say 'Let Amneh out of prison.'" Once freed, he was ordered to move from Salahudin to Sulaimaniya and set to work. But, in time, he came to suspect that Nuri was, in fact, an Iraqi agent intent on handing him over to Baghdad. He was therefore making the tape to alert the leadership of the Iraqi National Accord to what he perceived as the perfidy of their representative in Kurdistan.

The aim of the bombing campaign, by Amneh's account, was to impress Nuri's sponsors at the CIA with the capabilities of the organization they were funding. To that end, the agent was commissioned not only to organize the planting of bombs but also the distribution of leaflets in the streets of Baghdad. Handing out opposition propaganda in the heart of Saddam's capital would be a risky undertaking at the best of times, but the dangers were magnified by Nuri's insistence that the distribution be recorded on camera as proof that the leaflets had not simply been dumped. "Those leaflets," complained Amneh as he held up one such picture, "cost us more than a bomb. A bomb—somebody just takes it and leaves it. Leaflets need two people:

one to take photographs and the other to hand out the leaflets."

Despite such precautions, Amneh described Nuri as continually fretting that "the Americans will cut off financial aid to us." Whether or not Nuri's funding was curtailed, the burden of the bomber's complaint concerned the way his superior continually short changed him on pay and expenses. "We blew up a car and we were supposed to get two thousand dollars, but Adnan gave us one thousand," he grumbled at one point, going on to gripe that at a supply dump meant to contain two tons of explosives he had been given only a hundred pounds, the dump's custodian claiming that the rest had been stolen. He had not been able to buy a car or pay the dozen men on his team. On one occasion, Nuri had paid him with dollars that turned out to be counterfeit. Despite his position as a subcontractor for the richest intelligence agency in the world, he "had to buy clocks in the souk [market] and turn them into timers."

From the evidence of the tape, it appeared that the CIA was well aware of their agent's role in the Baghdad bombings and had even expressed some reservations. At one point, Amneh cited criticism from the Americans that he was "too much a terrorist," but observed that "Saddam Hussein has ruined the whole country, so how can anybody say we are terrorists?"

Rarely had a foot soldier in a covert operation been so voluntarily forthcoming about his work. Only once, he claimed, had he refused an assignment from Nuri. Soon after starting work, he had been asked to kill Ahmad Chalabi, leader of the other opposition group supported and funded by the CIA. Nuri had suggested using a booby-trapped car for the purpose, a proposal Amneh claimed to have declined on the grounds that this would make Chalabi a martyr—"You can say he is a thief, doesn't know how to work well, or mixes with the wrong kind of people, but none of this justifies killing him," and besides, "there will be Americans there."

Someone lacked his sense of moral discrimination. On October 31, 1995, a massive blast ripped apart one of the headquarters buildings used by the Iraqi National Congress in Salahudin. Twenty-eight people (though not Chalabi or any Americans) were killed, including the INC security chief. The CIA, as well as the INC and the KDP, all opened investigations. The Americans appropriated some bomb fragments from the scene of the blast, but refused to divulge any of their conclusions. The KDP, however, swiftly arrested three individuals who, after severe interrogation, eventually claimed that they were members of the Iraqi National Accord and had planted the bomb under orders from Adnan Nuri. Amneh repeated the accusation of Nuri's culpability on his tape.

Since the victims and the alleged perpetrators of this savage attack were both sponsored and subsidized by the CIA (no one suggested that the Americans themselves were behind it), it was hardly surprising that the agency remained mute on the results of its own investigation. The episode was even more embarrassing in view of the fact that the Accord was gaining favor among many senior agency officials as the more useful tool to deploy against Saddam even as their rivals for the CIA's affections lost favor.

The debacle of the INC's offensive in March 1995 had caused great irritation in powerful circles at Langley. Punishment was not long delayed. A month after the offensive, somebody leaked to the *New York Times* the previously secret fact that the INC was on the CIA payroll, thus removing Chalabi's fig leaf of respectable independence and thereby creating a furor in northern Iraq. In May, Chalabi was summoned to a meeting at the CIA station in London at which his high-level detractors in the agency planned to give him a savage dressing-down for launching the offensive without proper authorization. However, he still had friends and supporters in the agency, who proffered informed advice on how to deal with the angry bureaucrats. "Just tell them, 'I didn't do it and I won't do it again,'" counseled one of these pro-Chalabi partisans. "Bureaucrats in the CIA are used to dealing with their superiors, inferiors, people of the same rank, and bureaucrats in other agencies," this cynical veteran of covert operations later explained. "They are not used to dealing with someone like Ahmad who is capable of saying, 'Fuck you.' So the big meeting ended with Ahmad just getting a slap on the wrist."

Facing down his accusers may have been satisfying for Chalabi, but the chill from Washington only got colder. After the May encounter in London, orders were issued, apparently from the White House itself, that the INC leader was persona non grata at CIA headquarters. Since the agency was still sending teams to work with the INC in Salahudin, this seemed somewhat absurd, and Chalabi's supporters maneuvered to evade the ban on his visits.

Suspicions that the devastating attack on the INC in October had been carried out by agents of the Accord was a dramatic and extreme symptom of a widening split within the CIA itself. Increasingly, what had once been a relatively harmonious operation was splitting into two partisan groups, the devotees of Chalabi and the INC, and those who believed in the efficacy and glowing prospects of Iyad Alawi and the Iraqi National Accord. It was a phenomenon known as "clientism."

"Things got really bad," recalled Warren Marik, who had led the first CIA team to Salahudin and was very much in the Chalabi camp. "I realized that clientism was out of control when I saw some [Accord supporters] in the [CIA's] Iraq office challenge and scoff at the head of the office, who was only trying to keep a balance between the two sides. I suspected that they thought they could get away with it because they had a direct line to someone in the White House who was backing the 'zipless coup' idea."

The headquarters of the Accord was in London, which was why Iyad Alawi's supporters were concentrated in the CIA's London station, while those agency officers still loyal to Ahmad Chalabi and the INC were clustered in the Iraq operations office at Langley—though even there the lines were sharply divided. The battle that was fought in Kurdish mountain towns with bombs continued across the Atlantic, but now was fought with angry classified cables. On one occasion, for example, Iyad Alawi reported to his friends in the station that Chalabi had bounced a check. "London sent angry messages to Washington. 'Chalabi is bouncing checks, what has he done with all the money, this is a scandal, etc., etc.,'" recalled Marik. "I said, 'Hold on, let's see the check.' I called Ahmad, who said, 'It didn't bounce, I stopped payment because I didn't trust the vendor.' I checked it out, and it was indeed a stopped payment."

Despite such small victories, the balance of power was tilted against Chalabi. He was in bad odor because of the failed offensive and the eager enthusiasm from the upper levels of the agency for a quick solution. In any case, London had an additional advantage in the internal dispute because in the mid-1990s the station chiefs were influential men who had previously occupied very important posts.

"Basically," explained Marik, "the screwup was because of circumstances. Normally London is not so important. The station chief is usually some superannuated guy near retirement. But on this occasion, London was involved to a great extent because of the British involvement with Alawi. Second, it so happened that London was headed at that time by these eight-hundred-pound gorillas, Tom Twetton, an ex-DDO [deputy director of operations], and Jack Devine, an ex-acting DDO.

"So the whole thing got subsumed in vicious bureaucratic battles. Faxes flying back and forth about Ahmad's check, rather than thinking about getting rid of Saddam Hussein."

On the door frame at the Iraq operation's group offices at Langley, Bob Mattingly, who took over the group in late 1994, hung a banner illustrated with a quotation from a letter written by Winston Churchill after his appointment as head of the British Colonial Office, with responsibility for Iraq, in 1921. "I feel some misgivings," Churchill had declared, "about the political consequences to myself of taking on my shoulders the burden and odium of the Mesopotamia entanglement."

As internal relations inside the agency grew increasingly bitter and factionalized, a new and weighty factor appeared on the scene.

Coincidental with the INC's March 1995 offensive, President Clinton nominated John M. Deutch to be the director of the CIA. Woolsey, undermined by the fallout from his handling of the Ames affair, had resigned in January and two prospective replacements had successively self-destructed in the face of inquiries and revelations about their private lives. Deutch, the former provost of the Massachusetts Institute of Technology, was a highly intelligent man. He was also highly ambitious, having only reluctantly agreed to leave his powerful post of deputy secretary of defense, where he supervised the disbursement of over $250 billion a year, to take over the CIA. His eyes, according to many both inside and outside the agency, were firmly set on achieving a lifelong goal of becoming secretary of defense. An impressive performance at the CIA would aid him in that quest.

The new director was not universally popular among his staff, among whom his Pentagon ambitions were no secret. "There should be a rule against anyone running the CIA who wants to use it as a stepping-stone," one of his detractors from the agency sourly remarked.

If Deutch, who formally took over the CIA in May, did not impress senior subordinates, it is fair to say that the feeling was reciprocated. He made it clear that he had little regard for their skills and professionalism, and was in the habit of unfavorably comparing CIA officers to their faces with his former colleagues at the Pentagon. "He would be in a meeting with CIA people and if a military officer walked in, Deutch would say: 'At last we've got someone with brains here,'" asserted one former very senior CIA official. "Every finding that Deutch signed, he was thinking 'How does this improve my chances of becoming SecDef?'"

"Deutch mistrusted people, misunderstood things," recalled another detractor. "He was most in need of what he was least likely to seek—subordinates who would tell him his plans were a bad idea."

The new CIA chief had promised, on taking office, that he would "clean house" at the agency. Accordingly, new faces appeared in many key positions. Ted Price, who had done his best to hang on in the wake of the Ames disaster, finally left the office of deputy director for Operations and was replaced by David Cohen, who had spent most of his career on the intelligence analysis side of the agency. Like his master, Cohen did not inspire universal confidence among old hands in the operations directorate, where his only previous experience had been the humdrum task of debriefing Americans who had traveled abroad. The Near East Division, central to our story, was left in the hands of Steve Richter, a graduate of the agency's counterterrorism division, who had succeeded Frank Anderson in January. Richter was not highly regarded in all quarters, possibly because of ill feelings arising out of a dark and still highly secret episode back in 1988 when the CIA's entire spy network in Iran had been rounded up, with many subsequent executions. (An internal inquiry absolved Richter of any responsibility for the disaster.)

Most significantly of all, Deutch selected as his deputy director

the affable former congressional staffer and NSC intelligence director George Tenet, who had learned much about covert operations
without ever acquiring any direct experience. Thus the man who
while overseeing intelligence on the National Security Council staff
had consistently promoted the notion of a CIA-backed military coup
in Baghdad as a viable option was now high up in the direct chain of
command at the agency.

Several CIA officials formerly engaged on the Iraq operation
agree that Deutch's arrival at Langley coincided with a heightened
sense of urgency regarding the elimination of Saddam Hussein on
"the seventh floor," site of the office of the director. After reviewing
the record of the Iraq operation to date, Deutch's new management
team concluded that the Iraq operation should be made tighter and
more focused on the single objective of overthrowing the Iraqi
leader, without worrying about more general changes to the Iraqi
regime. Adopting procedures common to the management of Pentagon weapons programs, Deutch decreed "milestones," scheduled
points of progress toward the ultimate objective of the Iraqi leader's
downfall. If any of the newly promoted officials had doubts about
their master's eagerness to take on the burden of "the Mesopotamia
entanglement," they kept quiet. "Deutch recruited subordinates
who did not like to get yelled at," observes one retired official.

In truth, it seemed a propitious moment to push forward against
Saddam. The new team had hardly settled in at Langley at the
beginning of August 1995 when King Hussein sent his cautious
report of imminent and momentous developments in Iraq, followed
soon after by the dramatic news of Hussein Kamel's arrival in
Amman. While the agency soon wrote off Kamel as a potential asset,
the effect of his arrival on the position of the Jordanian government
was infinitely gratifying. The king swung decisively into the anti-
Saddam camp at last, and in consequence began to mend fences
with some important neighbors.

It had taken years for the Saudis to get over their pique at King
Hussein's soft line toward Saddam during the Gulf War, a position
dictated by the fervent support for Iraq among the majority of the
monarch's subjects. The Saudi attitude had been a source of great
frustration to the CIA, who wanted to be able to use Amman as a

base in plotting against Saddam, and Riyadh as a source of funds for the operation—the traditional Saudi role in its intelligence relationship with the United States. As one former senior CIA official recalled, "For years we were significantly affected by Saudi slowness in dealing with Jordan." Almost as soon as Kamel had arrived in Jordan, however, Prince Turki bin Feisel, the Saudi intelligence chief, made a "secret" but nonetheless widely noted trip to Amman to see the king, followed by a return trip to Riyadh by the anti-Iraqi Jordanian foreign minister, Abdul Karim Kabariti.

At the end of September, in the course of a trip to Washington, King Hussein and his foreign minister were invited out to Langley to be given a full-court briefing on the increasingly elaborate plans for a coup. Part of the briefing included an enthusiastic endorsement of Iyad Alawi's Iraqi National Accord. President Clinton himself, according to one of the king's advisers, pressed the royal visitor to give his full cooperation. Not everyone was so optimistic. An old friend of the king's, a man who had served in the CIA years before and knew the Middle East well, counseled caution. He later explained his instinctive reservations. "I wasn't given any briefings, but I'm like an old farmer who can smell bad weather coming. The people in charge [at the CIA] just weren't very experienced, but it's not the same outfit these days. This was not being professionally handled. Too many people knew about it—it was more of an overt operation than a covert operation. Still, the king didn't really take my advice. I guess when the president of the United States puts his arm around you and says, 'Your Majesty, we need your help,' it's hard to say no."

Matters were now moving forward at some speed. A new CIA station chief was dispatched to Jordan along with a special team devoted to the Iraq operation. These Americans were in turn serviced by a special unit created inside Jordanian intelligence, insulated from their colleagues, many of whom were suspected of being in the pay of the Iraqis. This special unit reported directly to the newly installed head of intelligence, Sami Batihki, and was responsible for assisting their CIA counterparts—interpreting for Arabic speakers, providing transport, facilitating secret meetings with Iraqi military officers, and other tasks. The CIA officer in day-to-day charge of the operation had previously served as an analyst in the agency's Directorate of Intelligence. Though lacking in any experience of covert operations, he was much esteemed at headquarters for the fluency and coherence with which he briefed superiors on the ongoing operation.

In mid-January 1996, David Cohen and Steve Richter flew to Riyadh for a grand conclave of high-level intelligence officials. Hosted by Prince Turki and the Saudi intelligence organization's desk officer for Iraq, General Abu Abdul Mohssan, those gathered around the table included the British MI-6, the Jordanians, and the Kuwaitis. By the time they adjourned the next day, it had been agreed that all would support the Iraqi National Accord in its forthcoming effort to displace Saddam. The Saudis, of course, had helped give birth to the Accord back in the distant days of 1990, only to see it split between the two founders, Iyad Alawi and Salih Omar Ali, allegedly because of a dispute over a check from Saudi intelligence. Alawi had moved to London, and there the Accord had risen again, now in the protective embrace of British intelligence. Those at the meeting also agreed on financial contributions for the undertaking. The Americans were authorizing $6 million, and the Saudis offered to contribute an equal sum. The Kuwaitis also made a pledge.

There had not been such aggressive and high-level interaction among the allies on the subject of Iraq since the Gulf War and, in retrospect, it is difficult to understand why. Certainly the defection of Hussein Kamel had energized the intelligence agencies of Saddam's enemies, but after the initial panic in Baghdad, the regime appeared to be firmly in control. There was, however, another factor at work of which CIA personnel were well aware, and that was the impending 1996 U.S. presidential election. According to former CIA officials closely engaged in the Iraq operation, pressure from on high to "move" against Saddam, which had increased from the moment Deutch took office in May 1995, became even more intense at the beginning of 1996. One such official stated to us that "It is my understanding that early in 1996 the CIA was given orders that 'You will mount a coup in this time frame. Just do it.' The orders came from the White House. Deutch signed off on it."

If there was indeed such a directive from the White House, it was very closely held. Political operatives dedicated to President Clinton's reelection attested later that they knew nothing of such a link between the election and the CIA covert action. "Not even the chief of staff would have known," said Harold Ickes, deputy chief of staff at the time. "It would have been between the President, Lake [National Security adviser], maybe Berger [Lake's deputy], and Deutch." Ickes added that he thought such a connection was "highly unlikely."

On the other hand, those who know Tony Lake, the deceptively mild-mannered and professorial National Security Council chief, suggest that, with a political payoff in mind, he would have been quite capable of approving such a bold initiative. "He was much more gung ho on this sort of thing than people might have thought," says one acquaintance.

Deutch himself denied that he received any such commission, a denial vehemently seconded by Lake. However, that was not the impression at Langley. As one former official recalled: "Deutch came back from a meeting at the White House, all fired up, stating words to the effect of 'Bring me the head of Saddam Hussein.'"

Politics aside, by early 1996, expectations in Washington, London, Amman, and Riyadh were certainly great that at long last the reign of the defiant Iraqi leader might be drawing to an abrupt close. The question therefore arises as to the reasons for such optimism. Iyad Alawi, as noted, had a remarkable gift for impressing officials in intelligence organizations. In addition, he had excellent relations with the regime in Amman. The king personally liked him, as did Abdul Kabariti. In contrast to his prejudicial comments about Chalabi—"smart but not wise"—Kabariti exhibited sincere admiration for Alawi, an indulgent attitude that was to survive the impending catastrophe.

"I met Alawi many times," Kabariti told us. "He impressed me in the way he analyzed internal Iraqi politics. He did not have high hopes for a coup against Saddam." If this was really Alawi's opinion at the time, he did not share it with the CIA contingent in Amman in the spring of 1996.

The Accord was by no means an impotent organization. Alawi and his immediate colleagues in the upper ranks of the organization had been influential figures in the Baathist regime. Salah al-Sheikhly, from a well-known Sunni religious family in Baghdad, was a statistician who had risen to a high position in the Iraqi central bank before defecting in the early 1980s. Tahseen Mu'alla had once enjoyed honored standing in the Baathist pantheon as the doctor who dressed Saddam's wound after the attempted assassination of Qassim. General Adnan Nuri had been a member of the Iraqi army's elite Special Forces and was married to the daughter of a close colleague of Hussein Kamel. Such men had an informed, if somewhat dated, understanding of the workings of Saddam's regime.

The second ingredient that appeared to bolster the prospects of the Accord was the opportunity, afforded by the change with the king's position, following Hussein Kamel's defection, to agree to mount operations out of Amman.

Ever since the Gulf War, the formerly sleepy Jordanian capital had taken on some of the atmosphere of Casablanca during World War II. Amman was the window through which Iraq and the outside world watched each other. High up on a hill in the suburb of Abdoun, the United States signaled its presence with a vast, newly constructed embassy, heavily fortified but also isolated from the chatter in the busy streets below. On Jebel Amman, a thoroughfare in the center of town, the Iraqi embassy constituted a diplomatic outpost second in importance only to the mission to the United Nations in New York, the ambassador carefully selected both for his abilities and loyalty.

The city was crowded with newcomers: Palestinians who had been expelled by the vengeful Kuwaitis immediately following the war; exiles from Iraq itself, many of them reduced to sleeping in mosques and public parks; businessmen growing rich in the smuggling trade; journalists who periodically flooded the hotels while importuning the somewhat sinister diplomats at the Iraqi embassy for visas to Baghdad. The arrival of the Palestinians and the richer exiles had sparked a building boom in the expanding city. Intelligence agents from across the Middle East and beyond congregated there, mingling with the agents of Saddam's Mukhabarat who also roamed the city, alert for threats to their master and occasionally murdering exiles, such as the nuclear physicist Muayad Hassan Naji, gunned down in the street in front of his wife and children in 1992 while on his way to Libya.

Of particular interest to the Accord and their American sponsors were officers of the Iraqi army who made their way out of Iraq and settled in Amman. Sometimes these were of exalted rank. In March 1996, for example, there arrived General Nizar al-Khazraji, former chief of staff of the Iraqi army and the man whose capture by rebels (though wounded, he had survived) had been the first indication of the seriousness of the southern uprising in 1991. The general, whom Saddam had tried to rescue from the rebels, now declared that "Saddam's policies have led to the fragmentation and breakdown of the unity of our land, our people, and our army." He announced his intention of working with the Accord and with "the devoted brothers in the military in Iraq."

However, al-Khazraji's defection was not greeted with quite the applause from Saddam's enemies that he may have expected. He was quickly interviewed by members of the CIA group riding herd on Alawi. They strongly urged him to place himself under the Accord leader's orders. "Why should I?" replied the former chief of staff. "I don't know this man." In consequence, he was soon left to his own devices, alone in a small house in Amman without even the necessary privilege of bodyguards.

Presumably, the CIA team felt they could afford to dispense with characters like al-Khazraji, who would have been greatly prized a few years earlier, because they were convinced that the Accord had already furnished them with links to a potent conspiracy against Saddam.

The first link in the chain was in Amman, in the form of a retired general from the Iraqi Special Forces helicopter force named Mohammed Abdullah al-Shawani. Al-Shawani, a native of the northern Iraqi city of Mosul, was living in the Jordanian capital but had not publicly broken connections with the regime in Baghdad.

In the fall of 1994, just before the first insertion of a CIA team into northern Iraq, al-Shawani came into contact with Iyad Alawi. He had a startling proposal: He and his three sons, Anmar, Ayead, and Atheer, were resolved to work to organize a coup against the leader. The young men were still living in Iraq and, furthermore, were officers, not merely in the army, but in the vaunted Republican Guard itself, where only recruits of impeccable political reliability were accepted. Anmar was a major, Ayead a captain, and Atheer a lieutenant. Known as staunch Baathists, they could circulate among their brother officers without attracting the immediate attention of the security services.

Alawi hastened to pass on this electrifying news to his friends in MI-6, who in turn shared it with the CIA. It was this development that ignited such enthusiasm among the coup enthusiasts in the London station, Langley, and the White House. By the end of the following year, the news from the al-Shawani brothers of the contacts they had forged in the Iraqi military and security system was sufficiently encouraging for the operation to go into high gear with the dispatch of the special CIA unit to Amman noted above. Once in Amman, however, the sponsors of the impending coup still had to communicate with their undercover allies in Baghdad, six hundred miles away, across the desert. It was to prove a fatal impediment.

Anyone wishing to send a message to the Iraqi capital had to send it in the care of one of the professional drivers sanctioned by the Iraqi Mukhabarat to make the regular run. Phone contact with Baghdad, at least since the war, had been difficult at the best of times. Since late in 1995 all international calls had to go via the telephone exchange at al-Rashedia, north of Baghdad, instead of being dialed directly. Operators taped all calls, and the recordings were subsequently examined by a special committee of representatives from the various intelligence agencies. For secret communications, the drivers were a vital and vulnerable link. Everything depended on their evading the pervasive scrutiny of Saddam's intelligence. But the Mukhabarat was well aware of the significance of the drivers, and devoted special care to watching their every move.

The CIA, once the team dedicated to assisting the coup effort had arrived in force in Amman, made it a priority to move beyond this professionally offensive system of hand-carried messages. The Accord was therefore furnished with a state-of-the-art satellite communication system, complete with high-technology encryption features to frustrate eavesdroppers. For further security, the Americans gave instruction on a system of code words and phrases to be used in conversation.

In the light of these painstaking security precautions, it was all the more astonishing that Iyad Alawi, in Amman to direct a secret conspiracy against the leader of one of the most efficiently repressive police states in the world, almost immediately began to broadcast both his presence and his intentions. On February 18, 1996, he held a press conference to announce the imminent opening of a headquarters in Amman. This event was, declared Alawi, a "historic moment in the Iraqi opposition movement . . . a beacon of light into Iraq, a light from which Saddam will find no hiding place." After venting particular spleen at Uday, who "profits from the black market, uses his gun freely on those who cannot defend themselves, physically abuses our women," Alawi did concede that there were areas of his organization's activities that "must remain secret if we are to succeed in our work and ensure that lives are not unnecessarily put at risk." If Saddam's security forces had previously failed to note the presence of Alawi and his attendants in Amman, they could hardly have remained in ignorance after this promotional exercise. Further announcements followed soon after, hailing the forthcoming launch of an Accord radio station in Jordan, al-Mustaqbal—"The Future." Alawi appeared on CNN (obligatory viewing in all Iraqi government offices), declaring that "We regard Jordan as the door to Iraq, and it is important for us to talk to people inside." In the same news item, the Jordanian information minister stated for the record that "We will not be involved in any plans to overthrow the regime. We think this should happen [sic] by the Iraqis peacefully."

A classified internal review by the State Department's Northern Gulf Affairs Bureau in mid-March concluded that U.S. policy toward Iraq was an "unqualified success."

On March 26, an array of Jordanian notables, together with the leading lights of the Accord, gathered to celebrate the inauguration of the new headquarters in a heavily guarded compound on the outskirts of Amman. The festivities were, however, marred by the appearance that morning in the London *Independent* of a front-page article by one of the present authors, Patrick Cockburn, describing the contents of the tape recorded by Abu Amneh two months before. Publication of the mercenary bomber's unedifying account of the Accord's terrorist campaign in Baghdad and the revelations of American sponsorship (not to mention the alleged role of the Accord in the slaughter at the INC headquarters in Salahudin), came as an embarrassing thunderbolt. Only two weeks before, President Clinton had hosted an "antiterror" conference at an Egyptian seaside resort to denounce terrorist bombings in Israel by the radical Palestinian group Hamas. Now the CIA was revealed as having indirectly sponsored similar tactics. (Following publication of the tape, Nuri hurried out of northern Iraq, eventually finding his way to Amman, where he quarreled with everyone before finally departing to sulk in Turkey.) But this unwelcome news was quickly followed by far more catastrophic information.

Sometime in January or February 1996, the inevitable had happened. A driver carrying messages from Amman to the coup plotters in Baghdad was intercepted and arrested. That would have been serious enough, but the man was carrying the vaunted high-technology satellite communication system donated by the CIA. At a stroke, the entire elaborately crafted plot—fiercely argued at the CIA, discussed at high-level intelligence conferences, discussed in the Oval Office, and possibly even factored into the campaign for the presidency of the United States—was brought to ruin.

With predictable craft and cunning, Saddam's intelligence officials chose not to give the slightest hint of their breakthrough. Instead, they waited, watched, and listened. The al-Shawani brothers, all unaware, faithfully followed instructions earlier communicated by their CIA advisers on evading surveillance and believed that they were still above suspicion.

It may well have been that the plot would have been blown even without the interception of the driver and his precious cargo. Several former CIA officials conveyed the view that the Accord was riddled with Iraqi double agents—"at least half" according to one official. Years later, reflecting on the disaster that overcame the scheme he had helped to foster, Prime Minister Kabariti concluded that the Accord's networks "were all penetrated by the Iraqi security service. The reason I think they were manipulated by Iraqi intelligence is that nothing succeeded, nothing worked."

But if the Accord's secrets were laid bare to Saddam's security services, there was also a leak from the inner recesses of the Mukhabarat.

Late in March 1996, just as Alawi was getting ready for the grand

opening of his new headquarters in Amman, a member of Iraqi security with whom Ahmad Chalabi's INC had previously been in contact
relayed an urgent dispatch to Salahudin. The Iraqis, he reported, had
the names of every single officer recruited by the Accord. Furthermore, the captured high-tech communications system was now
installed in and operated from an intelligence headquarters in Baghdad. The Iraqi intelligence officers, he reported, were thrilled at the
notion that they were communicating directly with CIA headquarters
in Langley. (In reality, the link was only with Amman.)

This was electrifying news, and at the end of March, Chalabi flew
to Washington. Ushered into the office of the director, he found
himself facing John Deutch and Steve Richter, head of the Near
East Division. The two men sat in silence as the INC leader
methodically presented the detailed evidence that their cherished
scheme was being brought to nought. It was an acid test of their
experience and professionalism. Would they accept that they had
been bested by the enemy, and retire gracefully from the field?

The answer was swift in coming. Consulted on the subject, the
young officers who had enthusiastically pushed forward when older
hands had counseled caution were unanimous in their rejection of
the unwelcome bulletin. Clearly, this was an exercise in spite by
Chalabi and his people, directed against their more successful rivals
for CIA funds and support. It was simply impossible, they argued,
for the Iraqis to have circumvented all their precautions against
penetration quite so successfully. If their plans had indeed been
compromised, argued the former intelligence analyst imported to
supervise the coup scheme, that was all the more reason to speed
things up. Deutch and Richter agreed. The operation would go on.

D day for the coup was set for the third week in June. So confident was Alawi in his prospects that he granted yet another interview, this time to the *Washington Post,* and without any apparent
intervention from his CIA case officers, in which he publicized the
forthcoming "secret" operation. With a lack of discretion that astonished the journalist, Alawi declared that the "uprising should have at
its very center the [Iraqi] armed forces. . . . We don't preach civil
war. On the contrary, we preach controlled, coordinated military
uprising [i.e., a coup], supported by the people, that would not allow

itself to go into acts of revenge or chaos." In other words, the Sunnis who had rallied to Saddam in 1991 out of fear of the Shia and the Kurds need not worry that this would be another intifada; there would be no "acts of revenge" against the regime's erstwhile supporters in the Baath Party. The interview was picked up and disseminated around the Middle East and the world by numerous wire services, with most of the stories emphasizing the connections among Alawi, the CIA, and the plans for an imminent coup.

If the interview was timed to kick off the coup, it may also have prompted Saddam to end his cat-and-mouse game. By Wednesday, June 26, arrests were in full swing. Later, the Accord came to believe that the arrests had begun earlier, possibly on June 20. Their scale and scope was a tribute both to the success of the coup plotters in spreading their net so widely across the Iraqi military and security apparatus and to the even greater success of Saddam's agents in monitoring the plot every step of the way.

The arrested officers—120 in the first sweep—were from the superelite Special Republican Guards, the General Security Service, the Republican Guard, and the regular army. They were all Sunnis, from Baghdad as well as key cities in the Sunni heartland that in the past had been staunchly loyal: Mosul, Tikrit, Faluja, and Ramadi. Some of the officers arrested were from a highly secret special communications unit called B32, attached to Saddam himself and responsible for his secure communications with military units around the country. So sensitive and important was the work of this unit that only those of unimpeachable loyalty had been accepted into its ranks. But even the B32's commander himself, Brigadier General Ata Samaw'al, was among those arrested, tortured, and executed.

This was the moment for Qusay, the quiet younger brother who had appeared so shy and deferential to Uday years before in the al-Rashid Hotel restaurant, to show his mettle. Saddam appointed him to head a special committee consisting of the heads of the Mukhabarat, General Security, and Military Intelligence. The committee was given unlimited powers to arrest any person, regardless of official status, who was implicated in the coup attempt.

Among those who passed into the hands of the committee, ensconced in the headquarters of the Mukhabarat in the up-market

Mansour district of Baghdad, were officers of very senior official status indeed. Apart from General Samaw'al, there was Colonel Omar al-Dhouri, a section director for the Amn al-Khass (the special security service), the most powerful of the intelligence services, and Colonel Riyadh al-Dhouri, from the Mukhabarat, both members of a tribe that in spite of disturbances the previous year was still considered loyal to the regime. A Tikriti general from the Amn al-Khass, Muwaffaq al-Nasiri, was picked up, as well as another from the Mukhabarat. Outside the military—where those implicated included several air force officers and two army generals—Qusay's scythe swept through the Dalaim clan, where several members of the leading families were arrested while others fled for their lives to Jordan.

Some of the victims had been even closer to Saddam and his family than the officers. The family's domestic staff was drawn from the small Assyrian Christian community of Iraq. They were now arrested and interrogated. Two cooks, Butrous Eliya Tome and William Matti, were later reported to have confessed to being involved in a plot to poison Saddam. Three months later, the number of those swept up in the purge had reportedly grown to eight hundred.

Needless to say, the three sons of Mohammed Abdullah al-Shawani, the young Republican Guard officers who had been at the heart of the plot, were among the first to be picked up. But they were not immediately executed. Qusay and his minions had other plans for them.

Sitting expectantly in Amman, the special CIA team was informed of the total and utter collapse of all their hopes in the most direct and brutal fashion. Their opponents in Baghdad could not resist the temptation of displaying the full extent of the Iraqi triumph. On June 26, the special communications device purred into use one last time, carrying a message from the Mukhabarat to the CIA. "We have arrested all your people," it reportedly said. "You might as well pack up and go home."

The CIA did just that. Within twenty-four hours, all the members of the group that had been working on the coup had left Amman. "They ran away," an embittered Iraqi exile said later. "Maybe they were scared, I don't know why." Some members of the Accord remained behind, issuing doleful press statements that chronicled the rout: "We have learned that several members of the special group [as the Accord termed the coup plotters] have died during interrogation. We mourn their deaths and promise them that their deaths will not be in vain."

When they sped out of Jordan, the CIA took with them General Mohammed Abdullah al-Shawani and lodged him in a safe house in London, the location of which was kept a closely guarded secret. A few weeks later, the safe-house phone rang. It was Anmar, the Republican Guard major and Mohammed Abdullah's eldest son, calling from Baghdad.

Anmar had a message for his father from his Iraqi captors. "If you are not in Baghdad in a week, Father," he said, "I and Ayead and Atheer will all be killed."

The old man broke down in tears. "What have I done, what have I done?" he reportedly cried. "I have killed my sons."

He did not go to Baghdad. No one imagined that the implied bargain would be honored. Instead, shaken by the contemptuous ease with which the Iraqis had once again penetrated their security, his protectors hurriedly moved the heartbroken father across the Atlantic.

The attempted Iraqi coup of 1996 marked one of the most colossal failures in the history of the CIA, deserving a place on the roster of such fiascos with the far more famous Cuban Bay of Pigs operation in 1961. So complete was the disaster that those concerned could only hope to evade condemnation by pretending nothing much out of the ordinary had occurred. "In the Central Intelligence Agency, like everywhere else in the world, they always have risk," said John Deutch later. "They aren't always successful. These were responsible risks carried out by dedicated individuals coordinated with an overall government policy." Asked whether he had understood beforehand that the coup plot had been penetrated (as Chalabi had warned him three months in advance), Deutch refused to comment.

To reinforce the notion that there was nothing for which it had to apologize, the CIA kept Alawi on the payroll, budgeting almost $5 million to support his activities in the following year alone.

In the meantime, Saddam, emboldened by his crushing victory, was turning his eyes north. There were fresh defeats and humiliations in store for his enemies.

TEN

Saddam Moves North

After long years of confrontation with Saddam Hussein, the U.S. government had gradually fallen into the habit of taking its most tangible asset for granted. Northern Iraq, the land of the Kurds, had been freed from Iraqi government control in 1991 only under pressure from Western public opinion, outraged by the spectacle of a million Kurdish refugees on the borders of Turkey and Iran. Thanks to George Bush's reluctant dispatch of allied troops into northern Iraq and the consequent withdrawal of the Iraqi military, the United States had acquired allies—the Kurdish groups—and a base from which to collect intelligence on the rest of Iraq. In addition, the fact that Saddam did not control a large portion of his own country was a valuable propaganda point. The CIA officials who first sat down at the end of May 1991 to ponder the future course of operations against Saddam had concluded that the existence of the northern safe haven gave them a public relations tool with which to "take a whack at his prestige," as one of them put it, "by accentuating his loss of sovereignty over the north."

By 1996, the U.S. presence in Kurdistan had taken on the appearance of permanence. U.S. aircraft patrolled the skies above the 36th parallel, a visible sign of U.S. protection as they enforced the northern "no-fly zone" for Iraqi aircraft. In Zakho, U.S. and allied officers staffed the Military Coordination Center, a relic of the 1991 negotiations that had led to the Iraqi withdrawal from Kurdistan and still provided a symbolic affirmation of Western military support. The State Department's Office of Foreign Disaster Assistance disbursed millions of dollars' worth of food and medicine annually. In Salahudin, the CIA teams continued to come and go, though since the debacle of the March 1995 offensive their role was strictly limited to collecting intelligence.

But this stability was entirely superficial. The two main Kurdish factions, Massoud Barzani's KDP and Jalal Talabani's PUK, had turned their guns on each other in 1994. The U.S. State Department had sponsored a cease-fire at meetings in Ireland in August and September 1995, but had lacked the interest or energy to push for a settlement on the underlying causes of the fighting. Barzani still refused to share the enormous revenues flowing into his coffers from the border crossing at Khabur, while Talabani declined to share power in Arbil, the administrative capital and also the largest city in Kurdistan, containing a fifth of the total population.

While Washington was playing less and less of an active role in Kurdistan, others were taking an increasingly disruptive interest. For Turkey and Iran, northern Iraq was an area of deepening concern. Since 1992, Turkey had been routinely sending military expeditions across the border in pursuit of Turkish Kurd guerrillas of the PKK. The Iranians had no love for the government in Baghdad, but neither did they want to see Iraqi rule permanently displaced by that of Turkey and the United States. Furthermore, they wanted to close off Iraqi Kurdistan as a safe haven for their own Kurds. Meanwhile, in Baghdad, Saddam watched the political currents in and around the northern provinces, waiting for the opportunity to reassert his power in his lost territories.

Underneath the umbrella of the two main parties in Kurdistan, there remained a multitude of smaller but significant power centers, especially tribes and clans such as the Harki, Zibari, and Sourchi, who still preserved a semifeudal social order amid the valleys and canyons of the fierce mountain landscape. The Sourchi, led by a rich and powerful family of that name that exercised authority over a dozen villages and a tribal army of several thousand men, not to mention business enterprises as far afield as London and Casablanca, were among the most powerful of these semi-independent entities.

High on a hilltop, the Sourchi home village of Kalaqin occupied a vital strategic position, for it overlooked the Hamilton Road, the highway built by a New Zealand engineer of that name in the 1920s to give the British access to the Kurdish heartlands. In a land almost without roads, it is *the* road. It runs from Arbil, the Kurdish capital, to Haj Omran, on the Iranian border, connecting the plains with the high mountains. Whoever controls the narrow, winding highway can cut Kurdistan in two, and warring armies have paid dearly to take or hold it. Even in a country almost continually at war for thirty-five years, the Hamilton Road has the reputation of being bathed in blood. Below Kalaqin, the road runs through the deep gorge of the Gali Ali Beg under towering black cliffs, where a single machine gunner can stop an army. As recently as the 1950s, the Sourchi were in the habit of periodically closing the gorge with the aid of a few bursts from the ancient Vickers machine gun on top of their mud fort, releasing traffic on the Hamilton Road only after suitable tolls had been extracted from stalled travelers.

By 1996, the masters of Kalaqin had long moved out of the fort and into luxurious villas in the family compound, but the Gali Ali Beg was as vital a strategic prize as ever, especially to Massoud Barzani and the KDP. It was the main supply route for their garrisons holding the front line in the intermittent civil war with Jalal Talabani's PUK, which held eastern Kurdistan. The tenuous cease-fire brokered by the Americans in Ireland still held, but the situation was fundamentally unstable. At all costs, Barzani had to protect his military lifeline.

The Sourchi had maintained an uneasy neutrality since the conclusion of the previous round of fighting between Barzani and Talabani in 1995 had left them in Barzani's territory. Now the KDP leader suspected that there was treachery afoot. His intelligence service had intercepted radio messages between Zayed, the eldest son of Hussein Agha al-Sourchi, the sixty-five-year-old head of the tribe, and PUK units to the east. The KDP later claimed that the messages contained military information, including details of the movements of Barzani, that might be of use to potential assassins.

Hoshyar Zibari, Barzani's veteran lieutenant, says that what happened next should not have surprised the Sourchi. He insists that the KDP demanded that "they either tell Zayed to go away, or at least hand over his radio." The Sourchi refused. They cannot have taken the demand too seriously, since they made no preparations to resist attack. They did not rally their sizable tribal militia, and Hussein Agha's large villa in the family compound was not fortified. The KDP did not have to move many men into the area. Because of its strategic importance, it already had detachments of Peshmerga nearby, notably at the old Iraqi army fortress of Spilik, across the valley on the other side of the Hamilton Road, to the east of Kalaqin. The KDP achieved complete surprise with its early morning attack.

"My father was expecting Massoud to come to lunch, not to attack him," says Zayed's brother Jahwar. "He was sleeping in his house protected by just three or four bodyguards when they attacked."

At 5:00 A.M., the KDP Peshmerga opened fire with Kalashnikov automatic rifles and rocket-propelled grenade launchers on Hussein Agha's villa. His bodyguards fired back. The attackers shouted at him to surrender, but he refused and fought on. By his family's account, the old man held out for over four hours against overwhelming odds. At the end, he climbed onto the flat roof of his house, presumably to shoot better. "He was hit by a rocket," says Jahwar. Wounded by the fragments, he was carried downstairs. When the KDP stormed the house, they shot him as he lay bleeding. Three of his bodyguards were also killed.

The KDP insist that they did not attack Kalaqin in order to kill Hussein Agha, but to arrest Zayed. They say the death of the Sourchi leader was "an unfortunate incident," and an accidental by-product of the attack. This explanation is belied by the fury with which all the houses in the village belonging to the Sourchi family were destroyed. Within a few days, demolition gangs had systematically leveled their villas, carefully taking care to remove the valuable reinforcement bars from the concrete. Ducks wandered through the wreckage of the sumptuous Sourchi homes, pitted with bullet holes, looking for food. A $3 million Sourchi-owned chicken farm near Kalaqin was dismantled and sold to Iran—"For peanuts!" laments Jahwar. Zayed himself, allegedly the object of the exercise, had escaped during the battle, and with other members of the family swore an eternal blood feud against Barzani.

Even in Kurdistan, many people found the killing of the venerable Sourchi chief rather shocking. It showed just how far Massoud Barzani was prepared to go in defending himself. "Many people talk about Massoud as a quiet and gentle person; but there is no gentle Barzani when it comes to his survival," says one Kurdish observer. "He knew the PUK was determined to finish him off." The speed with which Barzani reacted to what he saw as impending treachery by the Sourchi was a sign that he believed civil war was returning to northern Iraq. He was right, but this time the Kurds would not be left to fight on their own.

Jalal Talabani has always had the reputation of a gambler in Kurdish politics, being more mercurial than Massoud Barzani. Ever since he founded the Patriotic Union of Kurdistan in the wake of the great Kurdish defeat of 1975, Talabani has switched alliances with bewildering speed, even by local standards. In 1991, he was the first to kiss Saddam on the cheek—a gesture greeted with astonishment by other Kurds, not to mention his numerous friends in the West—but later he denounced negotiations with Baghdad.

In the uneasy calm that followed the Sourchi killing, Talabani was making final preparations for yet another dangerous gamble. He was planning to change the balance of power in Kurdistan and to do it with the aid of a major outside power: Iran.

Iranian support was essential for Talabani. Like other Kurdish leaders, he was not short of weapons such as Kalashnikovs or launchers for rocket-propelled grenades (RPG-7s). These were, in any case, part of the arsenal of any Kurdish household. He also had artillery, including multiple rocket launchers and 155-millimeter guns, but to use them he needed ammunition, which he could get only from Iran. The Iranians could also give him a military advantage over Barzani simply by allowing him to use the Iranian road system to ferry troops in safety up and down the long border on the

Iranian side. This enabled him to concentrate troops to attack the KDP wherever he wanted, outflanking their positions.

In return, Talabani could offer the Iranians his cooperation against the Iranian Kurds of the Kurdistan Democratic Party of Iran (KDPI). Documents captured by Barzani's forces later in 1996 showed Talabani cooperating with Iranian intelligence. Iranian Kurdish militants were arrested in his territory and were handed over to Iran. Very soon after the Sourchi killing, Talabani went further. Deep within his territory at Khoi Sanjaq, his own birthplace, the Iranian Kurdish guerrillas had a base, fortified with earth walls and machine-gun posts. In July, he agreed to allow Iran to send a column of two thousand Revolutionary Guards to capture it. In August, the Iranian Kurds signed an agreement with Talabani to stop all military operations against Iran. It was evidently the price for active Iranian support in the war that was about to start.

Barzani could see what was coming and began searching desperately for outside help of his own. Later, he was to make much of his warnings to Washington of the looming threat from Iran. His aides had, for example, faxed the official on the National Security Council who was responsible for the Middle East to report that Iran had "approached the KDP leadership on the evening of July 26–27 requesting access for their troops to come through Haj Omran, but Mr. Barzani refused to offer such access."

Washington may have imagined that the KDP was simply trying to win its support in a Kurdish faction fight by playing up the Iranian bogeyman and therefore there was no cause for alarm. This was a miscalculation that was to prove fatal to a large number of Iraqis in the very near future. "The chief American mistake," as Kamran Karadaghi, the highly astute Kurdish commentator, later observed, "was that they thought the Kurds had nowhere else to go."

On August 17, two months almost to the day after the incident at Kalaqin, the PUK launched its attack. It was cleverly timed to coincide with the fiftieth anniversary of the founding of the KDP in 1946, when the party leaders would be attending golden jubilee celebrations. KDP offices and checkpoints were decorated with the party's yellow flag and pictures of Mullah Mustafa, Massoud's father and the hero of the fight for Kurdish self-determination.

The first days' fighting were typical of warfare in the Kurdish mountains. The number of troops involved was not large, given the vast areas over which they were fighting. The PUK probably had, at most, seven to eight thousand trained Peshmerga and another five thousand militia. The KDP had similar numbers. Each side tried to hold villages, towns, strong points, and the few surfaced roads. It was a war of swift advances and retreats. Both sides tried to avoid heavy casualties to their hard-core units.

The early battles all went Talabani's way. KDP units at the northern end of the Hamilton Road swiftly crumbled because, so their commanders claimed, they were being attacked "with the help of Iranian artillery and rocket fire." Some KDP units changed sides. In the broken terrain, nobody on Barzani's side of the front really knew if the shell fire was coming from Talabani's forces or Iranian artillery forces firing across the border. Nevertheless, the KDP high command was insistent that their enemy was succeeding only with outside help. Hoshyar Zibari, Barzani's principal interlocutor with the outside world, said at the time that it was impossible to hold back the attack because it was "backed by howitzers and Katyusha rocket launchers provided by Iran."

Despite the outbreak of full-scale war in northern Iraq, the U.S. government gave no indication of concern or even awareness of what was happening. Only two months before, the CIA-backed coup organized from Amman, for which so much had been hoped and promised in Washington, had been routed with contemptuous ease by Saddam. Qusay's torture and execution squads were still mopping up the remnants of the conspiracy. President Clinton was in the midst of what seemed certain to be a triumphant reelection campaign in which the foreign policy of his administration barely featured as an issue. No one in the government wanted to raise the profile of Iraq at that particular moment.

The day that the PUK-Iranian onslaught fell on Barzani, he received a letter from Robert Pelletreau, assistant secretary of state for Near East Affairs, suggesting that he get together with Talabani for a peace meeting. Four days later, Barzani faxed Pelletreau with a plea for intervention: "We request the United States to . . . send a clear message to Iran to end its meddling in northern Iraq." The request was coupled with an ominous warning: "Our options are limited and since the U.S. is not responding even politically . . . the only option left is the Iraqis."

In fact, Barzani had apparently concluded that American promises of support, repeated many times by senior U.S. officials since 1991, were as worthless as similar promises had proved in his father's time. On August 22, hardly giving Pelletreau time to take action, Barzani composed a respectful request for help to the man who had killed three of his brothers, selectively murdered eight thousand of his tribe, and more generally slaughtered as many as two hundred thousand Kurds only eight years before. "His Excellency" Saddam Hussein was asked to "interfere to ease the foreign threat" from Iran.

Saddam was only too happy to oblige. He was already enjoying a very successful summer. In June, he had not only crushed the CIA-sponsored Accord coup, but his deputy prime minister, Tariq Aziz, had also deftly avoided a threatened American bombing attack. That threat had emerged as a result of attempts by Rolf Ekeus's Unscom inspectors to gain access to certain "sensitive sites" thought to contain information on Iraq's hidden weapons. The guards at the sites had been instructed to block the inspectors from entering; Unscom had taken the matter to the Security Council. The United States was highly confident that the Security Council would cite Iraq as being in "material breach" of the original cease-fire resolution, thus giving the Americans the authority to launch a military strike in retaliation. But Ekeus had flown to Baghdad, and, on June 22, to the intense irritation of the U.S. government, which thought he had conceded too much, had negotiated an agreement that ended the crisis and averted the strike.

Saddam, therefore, may well have felt that the international situation was shifting in his favor. When Barzani's request arrived two months later, the Iraqi leader was ready to take the risk and defy the Americans by interfering in the north.

Further letters from the KDP leader to Washington in the last week of August, warning again in tones of desperation that he might have to turn to Baghdad, were a smoke screen to conceal his real intentions. He had already made his arrangement with Saddam. Talabani's blitzkrieg was threatening KDP headquarters at

Salahudin on the mountain ridge overlooking the southern end of the Hamilton Road. Unless he received help soon, Barzani faced total defeat. The object of the exercise now, therefore, was to keep the Americans in ignorance of what was happening. Barzani therefore agreed to send emissaries to a meeting in the U.S. embassy in London on August 30. By this time, neither of the Kurdish leaders was interested in American mediation. Pelletreau later explained that he had telephoned Talabani to arrange a cease-fire, whereupon the PUK leader "promised full cooperation, [but] did nothing." According to the American diplomat's account, the United States was powerless to intervene. The State Department refused to come up with the money for a proposed mediation effort and the Pentagon wanted nothing whatsoever to do with northern Iraq.

On the ground around Arbil, the Kurdish capital and one of the oldest cities in the world, with a population of 600,000, the change in the political situation was more evident than in London and Washington. Since Iraqi forces had pulled out of most of Kurdistan in 1991, they had manned a heavily fortified line some fifteen to twenty miles from the city. The ground is flat and, apart from some earth ramparts, the Kurds had nothing to stop Iraqi armor.

Any resident of northern Iraq would have had good reason to dread the day when Iraqi tanks rolled out of those fortified lines and moved north. For one group in particular, however, the prospect was positively terrifying. Despite the vastly diminished standing of Ahmad Chalabi with his old patrons at the CIA, the Iraqi National Congress was still very much a presence in Kurdistan. The several thousand soldiers, administrators, intelligence officials, translators, broadcasters, and propagandists—many of them Iraqi Arabs without local Kurdish ties—who had remained faithful to the cause knew they could expect little mercy if Saddam returned.

On the other hand, the INC had achieved its moment of maximum success and popularity in 1994 when it had acted as a mediating force between the warring Kurdish groups. In London, where the United States was convening the peace talks due to begin on August 30, Ahmad Chalabi was pressing for support for a renewed INC mediation effort. It may well have been, given the scale of the fighting (and Talabani's confidence that he was winning), that the time for such efforts was past. But, in any case, such an effort was impossible without money. As Chalabi and anyone else familiar with Kurdistan well knew, a mediation force would have had to be able to grease its way with myriad payoffs to local commanders on either side to assure the safety, let alone the success, of the mediators. Nevertheless, there were officials in the State Department who thought this might be a good investment. Other more powerful government officials reportedly nixed the idea because of their antipathy to Ahmad Chalabi. But, sitting in London, Chalabi remained hopeful and therefore instructed his men in northern Iraq to hold themselves in readiness.

One of these INC leaders on the ground was the extremely astute and experienced Ahmad Allawi, who commanded an excellent intelligence service with spies in the Iraqi intelligence services and army. Soon after Talabani's bid to finish with the KDP had begun on August 17, Allawi says he "began to hear reports from Iraqi intelligence—both civil and military—saying they were preparing a huge attack on the north." He passed the news on to the INC headquarters in London.

Over the next week, Allawi found himself in a bizarre and ultimately tragic dilemma. On the one hand, there was the hope for a renewed INC mediation effort. If it was to play this role again, Allawi knew, the INC needed to concentrate its scattered forces. "We began to get them into big camps. They totaled between twenty-two and twenty-five hundred officers and men. We began to give them training and maps of Arbil."

Rumors of an Iraqi attack grew stronger. On August 29, Allawi sent out patrols behind the Iraqi lines to gather information. They returned with reports from their informants that the Iraqi army would begin to move soon. "On August thirtieth, we started to build defensive lines around Arbil," says Allawi. "At the same time, we got reports of a deal between the KDP and Baghdad. We told London."

Ghanim Jawad, senior official of the INC, who was at the INC headquarters in London, says that he and his colleagues were alarmed by what Allawi was telling them over the satellite phone from Arbil. Chalabi contacted the Americans with the news. But Barzani's delegates to the London talks were under instructions to allay suspicions of what was afoot. At the embassy meeting on August 30, an American diplomat asked one of the KDP delegates what was happening between his party and Iraq. "Nothing is happening," the delegate replied. "Everything is normal."

The CIA team in Salahudin at least was belatedly paying attention to the reports of Allawi and others. On August 27, acting on the basis of information from the INC that an Iraqi offensive was imminent, agency officers climbed into their vehicles and raced for the Turkish border. They had no intention of being anywhere near the front lines when Saddam made his move. Their allies in the Iraqi National Congress, fostered and funded by the agency from its inception, were left to fend for themselves.

The Iraqi attack began at 4:51 A.M. on Saturday, August 31, with heavy artillery fire from the east, west, and south of Arbil. The defenders saw some Iraqi helicopters. Half an hour later, Iraqi tanks began to roll forward against sporadic resistance from the INC militia, mostly former soldiers in the Iraqi army, and some three thousand Peshmerga of the PUK.

Directly in the path of the advancing Iraqi army was an INC camp at Qushtapa, just to the east of Arbil and about three miles from the Iraqi front line. It was not defended by any hills or natural features. The site had been chosen simply because it was close to the main road. A large unit of INC soldiers had been gathered here in a large, disused garage, awaiting orders from London to commence mediation efforts. Qushtapa was infamous in Kurdistan as the place Saddam Hussein had sent the women and children of the Barzani tribe, after he had massacred eight thousand male members of the tribe in 1983. Now it was to be the scene of another tragedy.

"The Iraqi army came straight across the fields," says Ghanim Jawad. "They surrounded the camp by eight or nine in the morning, collected the INC as prisoners, and put them in a big hall." Executions began immediately. An old woman who came to Qushtapa later in the afternoon of August 31 to look for her son was allowed into the forecourt of the camp by Iraqi soldiers. She said they had put the bodies of dead INC men into two open pits, in one of which she counted twenty-eight corpses. She said she could see fresh blood, showing that the killings had only just stopped. In all, ninety-six men were killed. Only six or seven escaped by putting on Kurdish Peshmerga uniforms, speaking Kurdish, and pretending to belong to the KDP. Ahmad Allawi says so few of his men got away because they were "caught in a sandwich and just couldn't escape."

As the Iraqi army was massacring the INC at Qushtapa, its tanks were moving into Arbil. Kosorat Rasul, the PUK commander in the city, could not defend it with only three thousand lightly armed troops against between thirty and forty thousand Iraqi soldiers. He had been extremely nervous about Iraqi troop movements since the previous day, repeatedly phoning Ahmad Allawi and Jalal Talabani, who was in his headquarters just outside Sulaimaniya, for information.

Talabani meanwhile was making frantic calls to Assistant Secretary of State Pelletreau to tell him that the Iraqis were coming and to plead for American intervention. Pelletreau responded with an assurance that there would be "serious consequences" if Saddam was indeed advancing into the north. The experienced diplomat was careful not to make any direct promises of American military intervention. But Talabani chose to interpret Pelletreau's judicious phrasing as meaning that U.S. help was on the way, or at least conveyed that message to his troops. On the front lines in front of Arbil, the defending troops waited expectantly for the first U.S. Air Force bombs to fall on the attackers.

The Iraqi advance into the city was slow and methodical. Saddam was also probably monitoring American reaction. The defenders were cheered when they heard American aircraft overhead at 10:40 A.M. in the morning. More appeared twenty minutes later, but the planes flew away and did not return. Allawi said the fight was hopeless: "We had only AK-47s and RPG-7s against tanks and Republican Guards. At two o'clock, the Iraqi tanks began to enter the city." During much of the morning, local PUK leaders, who had ruled Arbil for two years, were locked in prolonged discussions about what to do. Finally, Allawi told them that Iraqi tanks were in the center of town. The PUK leaders then issued their first decisive order of the day, which was that "everybody should escape as best they can."

By seven in the evening, the Iraqi flag was flying over what had been the Kurdish parliament, in the center of Arbil. Iraqi security quickly showed that they had a chillingly accurate knowledge of the whereabouts of their enemies in the city. Western diplomats and the Kurdish parties later decried the effectiveness of the INC, but Iraqi intelligence paid the opposition group a fatal compliment in the urgency with which they sought officers and members of the group. Nineteen of them were arrested by Iraqi security and taken to Baghdad, never to be seen again. Asked later if he knew what had happened to the people the CIA contingent had left behind as they fled to safety, Robert Pelletreau replied with chilling blandness: "You're asking whether some of them were killed? It is very possible that a lot of INC people were killed. But the INC's an independent organization."

Speaking from his campaign bus in Troy, Tennessee, on the day Arbil fell, President Clinton expressed "grave concern" about the situation but said it would be "highly premature to speculate on any response we might have."

Defense Secretary William Perry gave a hint that Saddam had little to fear from an American response when he said that U.S. vital interests were concentrated in the south and in the "strategic center" of Iraq, adding, "My judgment is that we should not be involved in the civil war in the north." Half a decade of American involvement in Iraqi Kurdistan was instantly forgotten.

American retaliation, when it did come on the second and third of September, provided a convincing demonstration of the limits of American power in the region. For the first time, formerly staunch Gulf War coalition allies Saudi Arabia and Turkey flatly refused to allow U.S. warplanes to attack Iraq from their territory. Clinton therefore unleashed unmanned cruise missiles from ships in the Persian Gulf, but the forty-four such missiles fired over two days were aimed at Iraqi command posts and air defense centers near Nassariyah, far to the south of the fighting. Iraqis, Kurds, and their immediate neighbors, if not the rest of the world, were very conscious that these targets were four hundred miles from Arbil. "They got the map of Iraq the wrong way up," remarked one INC official bitterly. (Among other excuses advanced by administration officials for steering clear of the north was the fear of being seen as an ally of Iran, Talabani's backer.) The United States also extended its southern no-fly zone, which had proved wholly ineffective in protecting the Iraqi Shia, seventy miles farther to the north, from the 32nd parallel to the 33rd parallel. "We have choked Saddam Hussein in the south," declared UN ambassador Madeleine Albright in an effort to put a gloss on the debacle. "We really whacked him."

It was left to CIA director John Deutch to pour cold water on official affirmations of success. Saddam, he bluntly informed the Senate Intelligence Committee on September 19, "is politically stronger now in the Middle East than he was before sending his troops into northern Iraq in recent weeks." Older and wiser perhaps after the failure of the attempted coup in June, Deutch, who may have known that Clinton was not going to make him secretary of defense, also stated that there was little prospect of Saddam being removed in the near term. This was in sharp contrast to the CIA's assessment to the same committee only four months before that "Saddam's prospects for surviving another year are declining."

Six years before, Saddam had foolishly stayed put in Kuwait while the Americans gathered up the will and the forces to punish him. He did not commit the same mistake again, withdrawing his military forces from Arbil almost as soon as the city had been secured. Nominally, the KDP was in control, its yellow flag replacing the green banner of the PUK on buildings throughout the city. But even as the tanks pulled back, Iraqi security remained behind.

The KDP, having made its Faustian bargain with Saddam Hussein, was eager to show that it could not be pushed too far. When the Iraqi Mukhabarat arrested some members of a small Kurdish Islamic group with whom the KDP enjoyed friendly relations, Barzani's men threatened to take some Iraqi intelligence men in Arbil and kill them unless the prisoners were returned. To their own amazement, the eight Islamic prisoners, who had in the interim been hung upside down and beaten with cables at intelligence headquarters in Mosul, found themselves driven back to Arbil and released. The KDP wanted to prove to the Kurds and the Americans that its deal with Saddam was "a limited agreement."

The subdued American response gave Saddam Hussein his first big political success since the invasion of Kuwait. He had calculated that the United States would not intervene, especially if he withdrew quickly, and he had been right. The fall of Arbil had no political repercussions in the United States. Despite pleas from his advisers,

Republican presidential candidate Robert Dole was loath to make it a campaign issue, conscious perhaps of the existence of a public record of his own fawning encounter with Saddam Hussein while on a visit to Iraq before the invasion of Kuwait.

The impact of the capture of Arbil was far greater in Iraq and the Middle East than in the United States or Europe. In Jordan, Prime Minister al-Kabariti remembers calling "the Americans." He said, "I don't want anything to do with you. It wasn't an embarrassment, what happened; it was something like treason."

With the fall of Arbil, the tide of battle in Kurdistan swiftly turned. Barzani's forces were now routing the PUK without any further aid from the Iraqis. Talabani's forces, additionally demoralized by their leader's foolish claim that the Iraqis were aiding the attackers with chemical weapons (he hoped thereby to provoke U.S. intervention), fell back in disarray toward the Iranian border.

Meanwhile, the surviving members of the INC in Kurdistan were running for their lives. None knew the extent of the cooperation between the KDP and the Iraqi Mukhabarat. Many of them made their way directly to Zakho, close to the Turkish border, but some two hundred and fifty were trapped in Salahudin in the al-Khadra Hotel, which had long served as their headquarters. Not far away, the house formerly inhabited by their CIA friends, who had often promised them that there was a detailed evacuation plan in the event of a disaster such as was now happening, stood empty and under KDP guard. There had indeed been an evacuation plan, but only for Americans. The INC members were desperate to leave but fearful that the KDP would hand them over to Baghdad before they reached Turkey.

During those tense days, the smell of fear and defeat in the hotel was almost tangible. Iraqis are often chain-smokers, but the survivors of the massacres in Arbil seemed to live from cigarette to cigarette. They sat in an office on overstuffed sofas, beneath a wall poster showing Saddam's famous victory arch of crossed sabers in Baghdad collapsing in ruin before the rising star of the INC.

"We expect death is coming," said Ahmed al-Nassari, one of their leaders. "There are Iraqi agents everywhere. We cannot abandon our weapons. The KDP is just an agent of Saddam." On September 15, there was a near riot as they milled about in the forecourt of the al-Khadra, clutching their submachine guns and waiting to board ten blue-and-white buses and two trucks, which were to drive them to Zakho. "If we hear nothing from the Kurds, we will simply go," said one of them in desperation.

For several nights they waited, their nerves cracking, for permission to leave. Ironically, the KDP may have delayed the exodus of the INC, as well as of the Kurds and Iraqis who had worked with foreign aid agencies, because they were a symbol of American involvement in Kurdistan. However much Barzani distrusted the United States, he wanted to keep the no-fly zone enforced by its warplanes as insurance against a full-scale Iraqi reoccupation. However, press reports of the plight of these people were surfacing in Washington, shaming the administration into at least putting pressure on the KDP to help them get to the border. An anonymous administration official told the *Washington Post* on September 9 that there would be no attempt to actually rescue the stranded INC members, excusing this morally questionable decision on the grounds that the CIA had "merely financed the group, not directed its activities inside Iraq." Furthermore, said the official, the CIA officers in Salahudin "had provided advance warning to the [Iraqi National] Congress of the Iraqi assault on Arbil, giving it ample time to flee." Reminded of this remark, Ahmad Chalabi observes that CIA officials "are not known for their veracity."

Finally, Karim Sinjari, the KDP head of security, arrived at the al-Khadra to say the INC could leave. They left behind about twenty of their Kurdish guards, who complained that they had not been paid for sixteen months. Their story turned out to be as sad as that of the INC. When Ahmad Chalabi had first come to Salahudin, he had asked the KDP for guards who were especially trustworthy, and was therefore given members of the Barzani tribe who had survived the massacre of 1983 because they had escaped or, in most cases, were children. "I lost my father and three uncles when they were taken with the eight thousand Barzanis in 1983," said Niyaz Salem as he stood in the abandoned INC headquarters. He spoke bitterly of the flight of the INC: "They took all the Arabs and left the Kurds, apart from a few. We feel betrayed. We don't know if they were CIA and we don't care."

Across Kurdistan the same scene was being repeated wherever

Kurds or Iraqis had worked with foreign agencies. On September 3, just as the first cruise missiles were being fired at Nassariyah, the Pentagon had ordered the evacuation of the allied Military Coordination Center at Zakho. The MCC had played a progressively reduced role since 1991, when the allies had first intervened in Kurdistan. Its activities were reduced further when U.S. aircraft accidentally shot down two U.S. helicopters in 1994, but it remained a symbol of allied protection of Kurdistan.

Within a few days of the evacuation of American, British, and other allied officers, the long, gray MCC building, sprouting aerials and satellite dishes from its roof, had become a holding station for anybody in northern Iraq associated with the United States. They were all visibly frightened. One man, speaking perfect American-accented English, said: "Iraqi law is extremely clear. Anybody who cooperates with foreigners is a traitor. When we heard that Saddam had offered an amnesty, we were even more frightened." Complimented on his English, he added bitterly: "Yes, I speak good English because I am one of those corrupted Kurds who deal with foreigners."

It was not just Kurds and Iraqis who had worked with American agencies who felt threatened. At Diyana, close to the northern end of the Hamilton Road, the Mines Advisory Group, a British charity, employed fifty Kurds to remove antipersonnel and antitank mines laid during the war along Iraq's border with Iran. It was dangerous work looking for aging mines with rusting trip wires in the undergrowth, but this was not what worried the fifty-odd men at the Diyana camp. When we visited them, they were nervously looking at an article in the September 12 issue of *Babel*, Uday's newspaper, which they passed from hand to hand. It contained a government statement spelling out the terms of an amnesty for Iraqi citizens who had worked for foreigners. The men in the camp were interested in a wide-ranging exclusion clause covering not only those guilty of murder, rape, and the theft of government property but also those who "spied for a foreign center." The Iraqi government's definition of espionage is notoriously elastic and the mine specialists had every reason to fear that it might be expanded to include them.

While they waited, the pendulum swung again in the Kurdish civil war. Barzani had swept all before him in his counterattack after the capture of Arbil and now the KDP leader was flushed with the triumph of what seemed like total victory. But Talabani's troops had retreated too fast to suffer heavy casualties. In the hidden valleys and mountain fastnesses along the Iranian border, they regrouped and prepared to counterattack in their turn. On October 13, reequipped by Iran, they swept out of their mountains and drove back the KDP. The flight of Barzani's men was as speedy as that of the PUK a month earlier. But the PUK stopped at the bridge at Degala, just before Arbil. Baghdad made it clear that it would use its tanks if Talabani tried to retake the city. There was a new referee in Kurdistan.

Saddam emerged as the clear winner from the Kurdish civil war of 1996. He had shown the limits of U.S. strength and resolve. He had eliminated Kurdistan as a safe haven for the CIA and its friends. The INC suffered a blow from which it would find it hard to recover. Some ninety-six of its members were executed at Qushtapa and another thirty-nine shot in Arbil itself, while between forty and fifty were killed in the fighting. It was a heavy loss for a small organization. With his victory, Saddam lifted the trade embargo he had enforced on Kurdistan since the end of 1991. Cheap Iraqi gasoline could now flow north unimpeded, as could the secret police.

By the end of September, the administration in Washington was moving to control the most visible evidence of the disaster in Kurdistan. Some 6,500 Iraqis and Kurds—members of the INC and their families as well as others who might have qualified under Uday's definition of espionage—were evacuated to the remote island of Guam in the northern Pacific. Here they were sequestered until the presidential election was safely over before being admitted to the United States.

Plucked from northern Iraq by the vagaries of war, politics, and covert action, these new immigrants might have thought that the worst of their problems were over when they reached American soil. For the vast majority, this was true. But others found themselves the victims of an extraordinary series of blunders by the FBI and the Immigration and Naturalization Service. It is difficult to imagine a clearer example of the callous ignorance habitually exhibited by the U.S. government toward Saddam's opponents and, indeed, to Iraq and the Middle East in general than the case of the six refugees who fled Saddam only to find themselves in an American jail.

While on Guam, the entire body of refugees was investigated by agents of the FBI assigned to ferret out any agents of Saddam, or other threats to U.S. national security, who might have infiltrated the group. The FBI agents were normally based in the United States, so to educate them in the intricacies of Iraqi and Kurdish politics, they were given a classified briefing by the CIA that lasted for forty-five minutes. Thus equipped, the agents set out for Guam and began their work.

One of the agents, Jennifer P. Rettig of the FBI's Chicago field office, grew suspicious while interviewing resistance fighter Hashim Qadir Hawlery. Hawlery asserted, in Arabic, that he had spent his life fighting in the "Kurdish liberation movement," a phrase the Egyptian-born U.S. Marine who had been pressed into service as an interpreter chose to shorten in translation to "KLM." Rettig had never heard of the KLM, and immediately deduced that it was a previously unknown and therefore highly suspect terrorist organization. In consequence, the unfortunate Hawlery was separated from his wife and seven children and consigned to the Los Angeles County jail for the next eighteen months while his lawyers fought immigration service efforts to deport him back to Saddam's Iraq and certain death.

Another of the cases concerned Major Safa al-Battat, who had deserted from the Iraqi army in 1991 and had gone on to become a hero of the Iraqi opposition. A native of Basra, he joined the INC and fought in the southern marshes against the Iraqi army. In 1994, he was poisoned with thallium, a poison commonly used against rats, when visiting Kurdistan. Thallium is favored by Iraqi security because it is very slow-acting, allowing the poisoner to get away before his victim dies. Al-Battat would certainly have expired had his friends not managed to get him to Britain, where he was successfully treated at a hospital in Cardiff. He could have stayed and enjoyed a peaceful life in the United Kingdom, but instead volunteered to return to northern Iraq where he became one of the INC's senior military commanders. In September 1996, he was airlifted to Guam, where he told his story to his official interviewer. The interviewer, however, concluded—possibly having confused thallium with Valium—that he was taking thallium for recreational purposes, as well as being an undercover Iraqi agent. At the time of this writing he was still in prison, appealing the decision to send him back to Iraq. Testimony in the case revealed that FBI agent Mark Merfalen believed the Iraqis "lie an awful lot" and that fellow agent John Cosenza believed that there is "no guilt in the Arab world, only shame."

As 1996 drew to a close, Saddam had reason to celebrate. He had scored three significant victories in the course of the year. In February, he had artfully lured Hussein Kamel back to his doom; in June, he had liquidated the most serious conspiracy against his rule to date and had discredited the Iraqi National Accord; in August, he had once more reasserted Iraqi influence in Kurdistan and had destroyed the INC, in the process exposing the weakness and indifference of American policy toward Iraq.

Yet, in the heart of Saddam's capital, a group of idealistic young people, unconnected with any known opposition group or foreign intelligence agency, were preparing to strike a dramatic blow against the ruling family. They had long had the will; recently they had found the means.

ELEVEN

Uday Takes a Hit

On a cool evening in December 1996, three identical white Mercedeses drove fast down Mansour Street, in West Baghdad. The road is long and straight as it passes the white walls of the old racetrack but ends abruptly at the T junction with International Street, forcing cars to slow as they approach the traffic lights. It was 7:25 P.M. and night was falling, but this is a wealthy neighborhood and the intersection is well-lit by street lamps. Anyone giving the convoy a close look might have noticed that all three vehicles bore identical license plates, a telling sign that somebody out of the ordinary was involved. A passerby brave enough to make a closer examination might have registered the identity of the passenger in the front seat of the leading car: Uday, the much-loathed eldest son of Saddam.

He had just come from feeding his pet police dogs, which he kept at the Jadriya Boat Club in South Baghdad, and was now on his way to a party being thrown by his friend and cousin Luai Khairallah. His bodyguards, used to Uday's obsessive socializing, were packed into the two following cars. The party was in a house only a few blocks away, and Mansour was an area Uday knew well and where he felt secure. It was filled with security men in uniform and plainclothes, most of them guarding the nearby Russian and Jordanian embassies or the haunts of senior Iraqi officials, like the Hunting Club, founded by Saddam himself in the late 1960s after he and his fellow Baathists found themselves blackballed from ancien régime establishments like the Mansour and Alwiya clubs.

As the cars neared the intersection, neither Uday nor the bodyguards had any cause to notice a young man with a sports bag at his feet standing nonchalantly near the Karkh Sports Club. He was eying the traffic coming down Mansour Street and had been there for several hours, but without attracting attention. Mansour is a fashionable district, full of small shops catering to the well-off Iraqis who live in the area, and the sidewalks are usually crowded with their customers.

The waiting man was not alone. Although they made no sign that they knew each other, he had three companions nearby, also with sports bags. Two lingered outside the busy Ruwad, a restaurant on the opposite corner of Mansour Street. A third stood by a Toyota pickup truck and a Toyota Super Salon sedan parked since earlier in the day on a nearby side street.

White Mercedeses are not common in Baghdad. The man outside the sports club had plenty of time to identify his target. As Uday's car approached, he reached down, unzipped his bag, and took out the Kalashnikov automatic rifle stowed inside. Unfolding the butt of the rifle, he stepped into the street and opened fire. His crucial role in the operation was to kill the driver of the lead Mercedes while the two men in front of the restaurant attacked the rest of the convoy. As he began to shoot, the others also produced AK-47s from their sports bags, each with four magazines holding thirty rounds of ammunition apiece, and closed in on the cars.

The lead driver was almost instantly torn apart by the hail of bullets. Everything appeared to be going as the attackers had planned. For months, they had been watching Uday as he roamed across Baghdad. Always, he took the wheel of the first car. They never discovered why, that night, December 12, he decided not to drive himself. Blazing away at the driver, the first gunman did not immediately realize that his real target was on the passenger side of the car. Seconds later, however, one of the men firing from outside the Ruwad saw that Uday, crouched down under the dashboard, was still unscathed. Shifting his aim, he poured the rest of his magazine into the most hated man in Iraq at almost point-blank range.

The gunmen had calculated that they would have a maximum of two minutes to start shooting, kill Uday, and make their escape. After ninety seconds, confident that they had accomplished their mission, the three men in the ambush party ran down Mansour Street and around the corner into the side street. The fourth gunman gave them covering fire to prevent the surviving bodyguards, most of whom were wounded or in a state of shock, from getting out of the Mercedeses and giving chase. The four men jumped into their two getaway vehicles, the pickup and the sedan, and sped away from the scene. Both the cars were stolen and had been fitted with false plates showing they came from Anbar, a province west of Baghdad famous for its loyalty to Saddam. The attackers figured that this made it less likely they would be stopped at checkpoints.

A diplomat from the nearby embassy of Jordan who arrived on the scene as the shooting ended recognized the bleeding Uday. The diplomat was convinced that he was dead after seeing him hit by the final, prolonged burst of gunfire from some nine feet away. Other eyewitnesses said he was covered in blood, but they could not tell how seriously he was injured. Uday was rushed to Ibn Sina Hospital, where Cuban doctors found he had been hit by eight bullets. Saddam himself soon arrived at the emergency room, where the Cubans finally were able to tell him, through a Spanish-speaking Iraqi woman, that his son would live. The translator said later that the Iraqi leader looked deeply relieved at the news.

Bad news about the health of the first family is normally a state secret in Iraq, but rumors were already spreading across Baghdad that Saddam himself had been seriously injured. So the media quickly reported the attack, stating that Uday had been "slightly wounded." The following morning, in a telling measure of Uday's grip on the Iraqi economy, the Baghdad stock market crashed as soon as it was announced that Uday's group of companies, which dominated the market, had ceased trading. The exchange rate for the dinar, always a reliable indicator of crisis in Iraq, plummeted to its lowest level in ten months.

In their lonely house in Tikrit, Raghad and Rina, the widows of

Hussein Kamel and his brother who lived in bitter isolation, celebrated the shooting of their brother, the man they regarded as primarily responsible for the downfall and murder of their husbands.

The government took time to admit the severity of Uday's wounds.
At first it was more interested in publicizing the fact that he was still
alive. Three days after the shooting, Baghdad radio announced that he
had telephoned the Iraqi national soccer team, competing in the Asian
games in the United Arab Emirates, "to bless the team's efforts." The
Iraqi Journalists Union, of which Uday was head, held a ceremony at
its headquarters "to celebrate that Uday Saddam Hussein had survived the sinful incident to which he was subjected on Thursday
evening." They slaughtered a sheep to show their "overwhelming joy"
that their chairman had survived.

On the following day, there was the first official Iraqi mention of
other casualties. The Iraqi media did this obliquely by announcing
that during a visit to the Ibn Sina Hospital Saddam had ordered that
those who were "seriously wounded in the cowardly attack" receive
the same medical care as his son. It did not say if they were bodyguards or passersby caught in the crossfire.

In the meantime, up to two thousand people had been arrested,
including hundreds of hapless shopkeepers and residents of Mansour. Even Sabawi and Watban, Saddam's half-brothers and bitter
enemies of Uday, were reportedly among those questioned. Watban
was still suffering from the wounds inflicted by Uday at the infamous party on the night that Hussein Kamel had fled to Jordan
almost eighteen months before. Uday himself, according to rumors
circulating among well-informed circles in Baghdad, later even
expressed the unworthy suspicion that his father might have had a
role in the assassination attempt.

This was the most serious blow ever struck at the ruling family,
but no one had any idea who was behind it, not that there was any
shortage of claimants for the credit. Al-Dawa, the venerable Shia
militant group that had existed since 1958, put out a statement in
Beirut saying it was responsible for trying to kill Uday. This boast
was not widely believed because al-Dawa had made few attacks in
Baghdad since the early 1980s. In addition, the group was known to
be very much under the control of Iran, whose rulers would be

unlikely to provoke Iraq by supporting a plot to kill the president's son. More credible was a claim from Kuwait by a member of the Dalaim tribe that the assassination was in revenge for the murder of General Mohammed Mazlum al-Dalaimi, tortured before his execution in 1995.

In the West, intelligence agencies were as baffled as Saddam by the dramatic shooting in Mansour. After having been seen running for their cars that night, the gunmen had utterly vanished, leaving only a haze of rumor and conjecture behind. Years of plotting by exiled opposition groups, $100 million in CIA money, not to mention the high-tech bombing efforts during the Gulf War, had all failed to inflict as much as a scratch on Saddam or a single member of his immediate family. Now someone had managed to locate and shoot Uday himself and then make a clean getaway.

Six months later, a fresh-faced Iraqi in his late twenties named Ismail Othman came to London and related to one of the present authors the real story of the attack and those behind it.

In 1991, following the chaos and ruin inflicted on Iraq after Saddam Hussein's Kuwaiti adventure, a group of well-educated young people in Baghdad founded an opposition group. They called it "al-Nahdah"—"The Awakening." Like other groups, these young idealists opposed the dictatorship, opposed the division of Iraq along racial and sectarian lines, and supported democracy. But there the resemblance to better-known political parties ended. Exile groups such as the Iraqi National Congress solicited publicity through conferences, interviews, and Web sites, attracting thereby not only funds from foreign intelligence agencies but also the unblinking scrutiny of Iraqi intelligence. Al-Nahdah remained a totally underground organization.

It did not have its origins in the Shia or Kurdish communities, whose activities the regime always carefully monitored. Most of its members were well educated, graduates from colleges in Baghdad. Many of them were women. The leader was Ali Hamoudi, an electrical engineer. His deputy was a woman named Raja Zangana, who had a job in the civil service.

"We studied how left-wing Latin American groups survived under repression by military dictatorships," explains Othman. The group organized itself into hermetically sealed cells in order to survive any

arrests—and the inevitable torture—among its members. It had so-
called "dead cells," which were inactive until needed to replace those
that were eliminated. Al-Nahdah was also careful to limit its communications with the world outside Iraq. Iraqi intelligence had a proven
record of success in intercepting links between opposition groups in
Baghdad and their headquarters in Kurdistan or Amman. Therefore,
Ali Hamoudi, the secretary general, instituted a policy under which
any member of al-Nahdah who traveled abroad was automatically isolated from the rest of the organization. At one point, Jordanian intelligence heard rumors that such a group existed, but failed to penetrate
or even find it. "Their security was very, very good," says one of the few
outsiders who ever came to know them.

Early on, they considered emulating the Latin Americans and
launching an armed struggle against the regime. But by 1994, according to Othman, the group decided that while they were not strong
enough to launch regular guerrilla warfare, they could carry out selec-
tive assassinations. "We thought the regime had four pillars," says Othman. "Saddam himself, Uday, Uday's younger brother, Qusay, and
their cousin Ali Hassan al-Majid."

They considered the option of assassinating Saddam Hussein, but
quickly concluded that this was impossible owing to the care with
which the Iraqi leader concealed himself. They were aware that even
his senior ministers did not know where he was at any particular time.
Uday, on the other hand, was a more viable target because of his hec-
tic social life and frequent business meetings. Al-Nahdah also thought
that, aside from Saddam himself, he was the leader whose elimination
would do most to destabilize the regime. "After Saddam, Uday had the
most authority," says Othman. "He would often make decisions without consulting his father. We decided to kill Uday." Members of the
group began to track his movements in Baghdad.

Their first attempt came in April 1966. Al-Nahdah believed it had
good information that Uday would visit a farm he owned in Salman
Pak, an hour's drive southeast of Baghdad. They activated a military
cell and waited for him, but he failed to turn up. The group faced the
same problem as other Iraqis who had considered killing senior members of the regime over the previous thirty years. "There have always
been Shia willing to die to assassinate leading members of the regime,"

says one Iraqi intellectual. "But they never had access to the intelligence you would need to be successful."

A month after the abortive Salman Pak attack, al-Nahdah had its first serious reverse, which might have destroyed the group had it not been for its system of cutouts. Secretary General Ali Hamoudi was arrested at a house where he was hiding in Saddam City (al-Thawra), the great Shia slum where almost half of the population of Baghdad lives. When Raja Zangana, his deputy, came to visit him, she was also detained. Othman says: "They didn't succeed in getting any information out of him. He died under torture. She was executed at the end of September and her body handed back to her brother in early October." The cell structure of al-Nahdah prevented Iraqi security from unraveling its organization. A member whose identity they feared might be revealed by their imprisoned leaders was sent out of the country.

Precise intelligence about the movements of Saddam's inner circle could come only from within the elite itself. The flight to Amman of Hussein and Saddam Kamel showed how difficult it was for dissident members of his family to cooperate with members of an opposition who had dedicated their lives to overthrowing the regime. The breakthrough for al-Nahdah came only because of a blood feud within Saddam's family, which had caused one member to vow revenge against his own clan.

In the last months of 1996, al-Nahdah came into contact with Ra'ad al-Hazaa, a Tikriti and a relative of Saddam's. Up until 1990, al-Hazaa had been a trusted member of the presidential guard. At that point, his career was blighted by the independent actions of a member of his family—a fate suffered by many in Saddam's Iraq. His uncle, General Omar al-Hazaa, had formerly been a divisional commander in the Iraqi army but had retired soon after the outbreak of the war with Iran. Thereafter, the general spent much of his time at a comfortable officers' club near his home in the Yarmuk district of Baghdad.

According to another Iraqi officer, now in exile, the general was known for drinking heavily at his club and, when drunk, would often criticize Saddam for his conduct of the war. The consequences were inevitable. "In 1990, the general was arrested," the exiled officer relates. "He was taken to al-Ouija and his tongue was cut out. Then he was executed. His son Farouq was killed at the same time, and the general's house in Baghdad was bulldozed."

Saddam later showed himself uneasy at the savage punishments inflicted on his al-Hazaa relatives. At a family gathering around Uday's bedside after he was shot, the Iraqi leader distanced himself from what had happened in 1990. He blamed the executions on Ali Hassan al-Majid and Hussein Kamel. By now, the latter had been dead for a year, but Ali Hassan was present and Saddam was scathing about his actions. "It was you and Hussein Kamel," fumed the dictator, "who caused me to execute Omar al-Hazaa and his son, and had it not been for your persistence and provocation, I would not have embarked on that action. Together you pursued members of his family, and their houses were destroyed on your orders. It will always be said that Saddam did that. People will not say that Ali Hassan or Hussein Kamel did it."

Although he lost his job as a presidential guard, Ra'ad al-Hazaa survived the death and disgrace of his uncle. He even remained a habitué of the social circle around Saddam's inner family. Most important, he was still a friend of Uday's relative and boon companion Luai Khairallah. By the end of 1996, unknown to the rest of his family, he was in contact with members of al-Nahdah, who realized that he could provide the critical intelligence they needed.

On December 9, 1996, Ra'ad was having a drink at Luai's house when his host let slip a vital piece of information. "We are planning a party in Mansour on Thursday," he said, and issued an invitation. While giving the address, Luai mentioned that Uday was going to be there. Ra'ad immediately passed the news on to his contact with al-Nahdah.

"We gave the information to our group to be ready in three days' time," says Othman. "We knew the route Uday was likely to take to the party." They chose the intersection of Mansour and International streets as the perfect spot to lie in wait because in driving to the party Uday would have to pass that way, no matter what direction he was coming from. The long, straight-approach road meant they could see the car coming from some way off and then spring the ambush.

Uday did not die, but al-Nahdah nonetheless considered their bold attack a success. "We proved that the Iraqi people could still act after the crushing of the uprising in 1991," says Othman. "We wanted to end the common sense of hopelessness. The Iraqi opposition had all gone out of the country, with nobody left inside." The group also knew that the political damage to the regime would be greater if the assassins could escape undetected and unscathed.

There was chaos in the area in the minutes immediately following the attack as the security forces frantically reacted to what had happened. The main roads were sealed off, but the gunmen had already gone. Othman's story is that they drove west and took refuge with a bedouin tribe for four days. There they were joined by Ra'ad al-Hazaa, after which they all made their way to Jordan. He says that they chose this route because they knew Iraqi security would expect them to escape to Iraqi Kurdistan or Iran, each of which is less than three hours' drive from Baghdad.

It is not a likely story. Western Iraq is largely uninhabited desert, and its tribes would be unlikely to give sanctuary to men they did not know at a time when Iraqi security was scouring the country looking for Uday's attackers. Nor would Jordan, whose security service is thoroughly infiltrated by Iraq, be a safe refuge. Al-Nahdah had considered sending the gunmen to Kurdistan but "we were advised not to go because Saddam's agents were very active there since the invasion of Arbil." Instead, according to unimpeachable sources, al-Hazaa and the gunmen took the obvious course and escaped over the Iranian border.

Once in Iran, their problems were not over. The Iraqi government was officially demanding their return. They feared that Iranian security, in an undercover deal with Baghdad, might hand them back. They therefore got in touch with Sayid Majid al-Khoie, the leading Shia clergyman from Najaf who was in exile in London. Explaining what they had done, they asked him to intervene with the authorities in Tehran. They also wanted to make sure that if they could not stay in Iran, they would be allowed to go to a third country where they would be safe. The cleric persuaded the Iranians to cooperate, although the authorities in Tehran insisted that the al-Nahdah members avoid any mention of their escape to Iran (hence the misinformation about the escape to Jordan). Eventually the gunmen moved on to Afghanistan, posing as Iranians, though their neighbors in Kabul were perplexed about the origin of "the young men who arrived in their street speaking such poor Farsi."

The core of the organization remained in Baghdad. A former Baathist leader, now in exile, says that Saddam appointed three different investigations into who was behind the December 12 attack, but without success. A sign of their failure is that almost two years later, in August 1998, Iraqi security announced that it had arrested a dozen people involved in the attempted assassination. All indications are that those arrested had nothing to do with the attack on Uday.

The heaviest blow to fall on al-Nahdah happened entirely by accident. On February 2, ten weeks after the attack on Uday, some of its members gathered for a meeting at a house in al-Kreeat, a pleasant suburb in North Baghdad. It is full of trees and well known for its market gardens and busy restaurants lining the bank of the Tigris. Suddenly, one of the al-Nahdah guards saw a soldier climbing over the fence surrounding the house. He immediately opened fire. A gun battle broke out, which went on for four hours. "They used rocket-propelled grenades and destroyed the house over the heads of our people," says Othman. "Eleven people in the house were killed along with an officer and two soldiers."

The security forces had arrived on the scene not because they had tracked down al-Nahdah, but through sheer chance. One of the organization's members had recently bought a car, not realizing that it was stolen and its documents forged. On a routine check, the police discovered the stolen car outside the house in al-Kreeat and realized that some sort of meeting was going on inside (always grounds for official suspicion in Iraq). Othman's list of those who died in al-Kreeat confirms the impression that most members of al-Nahdah are well-educated professionals: Ra'ad Kamil, a pharmacist; Saif Nuri Mohammed, a goldsmith; several others who worked in the planning and education ministries.

In the space of just over a year, Saddam had seen two of his sons-in-law killed, his half-brother shot in the leg, and his eldest son riddled with bullets. Even if he was scoring significant successes against the Americans at Arbil and elsewhere, this was clearly not a happy family. Early in 1997, he summoned surviving members to gather around Uday's bed in the Ibn Sina Hospital for an extraordinary family meeting. All the pillars of the regime al-Nahdah wanted to eliminate were there—Qusay, Ali Hassan al-Majid, Saddam's two half-brothers Watban and Sabawi, as well as the recumbent Uday. Saddam may have always intended the tape of what he said to be made public (it found its way to London), since he systematically blames his relatives for many acts of violence and corruption in Iraq previously attributed to himself. He also tells them that they owe everything to him, inheriting "power, influence, and standing, which you are using in the ugliest way . . . we are not a monarchy, at least not yet."

Saddam begins by reminding Ali Hassan al-Majid that before the 1968 revolution "you were a lance corporal and a driver in Kirkuk." He says one of the reasons he dismissed him as defense minister in 1995 was because he was smuggling grain to Iran. After throwing in the recriminations over the killing of General al-Hazaa, Saddam moves on. The performance in office of his half-brothers Watban and Sabawi is treated with scornful contempt. "You, Watban, must know that the Interior Ministry was ruined during your term," says Saddam. "As for Sabawi, what kind of a security director is he, in a country experiencing such conditions? He goes to his office at 1100 hours, half asleep. He left public security to dishonest elements who are stealing people's money. I had to execute some of them."

The diatribe continues with a reference to Luai Khairallah, who (unknown to Saddam) had accidentally given the al-Nahdah assassins their chance to intercept Uday. Luai had reached agreements "with Mafia and drug traders to smuggle sums of money to be laundered in Iraq." He then accuses Qusay of being two-faced, but much of his bile is reserved for Uday: "Your behavior, Uday, is bad, and there could be no worse behavior than yours. . . . We want to know what kind of person you are," says the father. "Are you a politician, a trader, a people's leader, or a playboy? You must know that you have done nothing for this homeland or this people. The opposite is true."

By the time Saddam spoke, it was evident that Uday was too seriously wounded to return to his former role as his father's viceroy.

The Information Ministry admitted he had been hit by eight bullets. The government tried and failed to send him to France for treatment. Although by 1998 he was driving again, he had almost entirely lost the use of one leg. Iraqis hopefully circulated a rumor that he had become impotent. According to his former friend and editor Abbas Jenabi, who fled Iraq in September 1998, this was far from the case. Uday continued to have sex with as many as four different women a day, some of them as young as eleven. He had certainly not lost his zest for business and was still active in a multitude of profitable sanctions-busting smuggling operations, particularly through his Asia and Kani companies.

More important, and despite Saddam's bedside strictures, Uday gradually began to meddle in politics again, performing his usual role of disrupting relations in the inner family. By 1998, he was on the offensive, this time taking aim at his uncle Barzan, still serving as Iraqi ambassador to the UN in Geneva. Officials connected with Barzan came under attack in Uday's newspaper, which also placed renewed emphasis on the close relations between the proprietor and Saddam— "the beloved apple of his father's eye . . . the lion's eldest cub." On August 30, 1998, Barzan was recalled to Baghdad. At first he refused to leave Switzerland. A replacement ambassador arrived, Khalid Hussein, formerly Uday's office director at the Olympic committee. After initially hinting that he was resigning from the Iraqi foreign service and remaining in Switzerland as a private citizen, Barzan returned to Baghdad.

Despite Uday's survival and eventual resurgence, al-Nahdah, without money or resources, succeeded in doing more damage to the regime than the Iraqi National Congress and the Iraqi National Accord combined. It showed that the family was vulnerable, and had thereby damaged, if not destroyed, the aura of invincibility that surrounded Saddam and his immediate kin. But in one sense the attempted assassination of Uday came too late. The killing of Hussein Kamel, the defeat of the INA's conspiracy, and the entry of Iraqi tanks into Arbil had made Saddam stronger than at any other time since 1991. The Iraqi leader was preparing to go on the offensive.

TWELVE

Endgame

In her four years as U.S. ambassador to the United Nations, Madeleine Albright had staked a claim to be regarded as Saddam Hussein's most unremitting foe. Her answer to the question posed in a 1996 TV interview regarding the cost of sanctions in Iraqi childrens' lives—"We think the price is worth it"—which became famous in the Arab world, only underscored her hawkish credentials at home on the issue and did nothing to impede her eventual elevation as secretary of state.

Soon after Mrs. Albright's arrival in Washington, word spread that she would be making a major policy address on the subject of Iraq at Georgetown University. Expectations ran high, on all sides. Before the speech, a prominent businessman of Iraqi extraction, known to be in close touch with Nizar Hamdoon, Saddam's UN envoy, was circulating word among the Iraqi exile community in Washington that the speech would contain dramatic new initiatives.

On the appointed day, March 26, 1997, Mrs. Albright strode onto the dais and announced that "We do not agree with the nations who argue that if Iraq complies with its obligations concerning weapons of mass destruction, sanctions should be lifted." Sanctions, she made clear, would remain. Almost six years had passed since Robert

M. Gates, President Bush's deputy National Security adviser, had declared that sanctions would remain as long as Saddam Hussein ruled Iraq, and that in the meantime "Iraqis will pay the price." Nothing, it seemed, had changed.

There could have been no clearer message to Saddam that he had little to gain in further cooperation with the UN inspectors. Even had he been of a mind to yield the secrets of the weapons he had so tenaciously concealed since 1991, Albright had told the world that he would gain nothing by doing so.

Yet while stating that Saddam's putative arsenal of mass-destruction weapons was unconnected to the maintenance of sanctions, the United States still emphasized the importance of the weapons inspectors' mission, the execution of which, paradoxically, depended on Iraqi cooperation and assistance. It was up to the Iraqis to escort inspectors to sites where weapons or documents might be hidden. The Iraqis had repeatedly demonstrated their power to exclude the inspectors from any site if they so wished. The extensive program of remote cameras and other sensors monitoring former weapons-related factories and laboratories could be removed with a simple phone call from Baghdad. In that event, the only remaining sanction for the United States and its allies would be military action—a renewed bombing offensive. But the threat of force was a diminishing asset because, as the failure to secure support for bombing Saddam in retaliation for the Arbil operation had vividly demonstrated, every potential military confrontation highlighted declining support for the United States both in the Middle East and around the world. By 1997, recalls a senior Unscom official, Security Council resolutions condemning Iraq had "all the impact of traffic tickets." Thus, a crisis over Iraqi cooperation with Unscom carried significant risks for the United States. Saddam, as he well recognized, could choose the timing of those crises. As 1997 went on, he had plenty of opportunity to do so. The inspectors were testing the limits of the Iraqi leader's patience.

Ever since the inspectors first arrived, Saddam had been forced to give up much. His initial expectation that his Unscom problem would last only a few months and that the inspectors could easily be fooled or bribed had soon been proved false, as we have seen. Thereafter, the

Iraqis had waged a fighting retreat. Up until the summer of 1995, they had successfully concealed their most modern chemical capabilities—the VX nerve agent—as well as their homegrown missile program and almost the entire biological effort. Then, in August 1995, the defection of Hussein Kamel had brought disaster. Unscom officials now knew that they had been successfully fooled by their Iraqi opponents. The inspectors set to work to uncover the full truth.

In April 1997, Rolf Ekeus reported to the Security Council that after six years of work, "Not much is unknown about Iraq's retained proscribed weapons capabilities. However, what is still not accounted for cannot be neglected." Even a few long-range missiles, he wrote, would be a source of deep concern if those missiles were fitted with the most deadly of chemical nerve agents, VX. "A single missile warhead filled with the biological warfare agent anthrax could spread many millions of lethal doses in an attack on any city" in the Middle East.

Ironically, publicity about Saddam's secret arsenal, attendant on Ekeus's investigations, helped him project a chill of fear over his neighbors. The primary effectiveness of these weapons was psychological. In 1991, the Kurds had thought they were under chemical attack and had fled in panic when Saddam's troops dropped flour on them from helicopters. "I lie awake at night worrying about those terrible biological weapons," a tremulous King Fahd of Saudi Arabia once told a visiting Kuwaiti diplomat. To divest Saddam of the psychological advantage he derived from his tiny but famous arsenal, Unscom would have to find or account for every single missile, all the VX, and every pound of anthrax, as well as the machines and materials used to make them. That was almost certainly an impossible undertaking, but even in making the attempt, the inspectors would have to penetrate and defeat the system of concealment created on Saddam's orders in the early summer of 1991.

As with so much else, the existence of this system had come to light thanks to Hussein Kamel. The sudden appearance of the huge cache of "chicken farm" documents together with the fact, soon deduced by the Unscom sleuths, that certain categories of files that would naturally belong in such a collection of records were absent led them to the inescapable conclusion that the missing documents must still exist under the protective guard of a concealment apparatus dedicated to frustrating Unscom. Hussein Kamel's cousin and fellow defector, Major Izz al-Din al-Majid of the Special Republican Guard, who had actually had missile parts buried in his own garden in Baghdad, provided confirmation and a wealth of detail in interviews with Unscom officials.

As we have seen, concealment was in the hands of especially trusted members of elite security organizations: the Mukhabarat, the Special Republican Guards, and the Special Security Service. Once upon a time, this arrangement had operated under the supervision of Hussein Kamel, but after his departure, control had passed to the capable and hardworking Qusay, operating in conjunction with the immensely powerful Abed Hamid Mahmoud, Saddam's private secretary. Not everyone, of course, in the twenty-thousand-man Special Republican Guards, and the Special Security Organization, which comprised a total of two thousand people, was involved in the exercise. Those directly concerned numbered no more than a few hundred, selected on the basis of absolutely unquestioned loyalty and, usually, a direct family relationship with the leader.

At the end of 1995, Ekeus commissioned Nikita Smidovich, the mustachioed Russian expert who so maddened senior Iraqi officials, to begin leading an inspection team targeted specifically on the "mechanism" for concealing missile parts, tools, and, most important, documents. Since the weapons were being guarded by the same security organizations that protected Saddam Hussein himself, that meant Smidovich and his team would inevitably be getting very close to the central nervous system of the regime itself. In March and June 1996, Smidovich had tried to get into what became known as "sensitive sites" occupied by these security organizations and had been blocked, or at least delayed, by the guards. Ekeus managed to hammer out a compromise in June 1996 with Tariq Aziz under which the teams would be allowed into such places. But the following month, the team was blocked again at a Special Republican Guard camp, although they saw long, round objects looking for all the world like Scud missiles being hurriedly driven away. The Iraqis had a ready explanation: The admittedly suspicious "Scud-like objects" being removed from the site were, they claimed, concrete pillars that coincidentally resembled missiles.

As the Unscom teams continued their hunt, they found that time and again they were just too late. Despite stringent efforts to make their descent on a suspected site a total surprise, the Iraqis appeared to have been forewarned in the nick of time and the team would arrive to see trucks speeding away in the opposite direction. Either Iraqi intelligence had managed to find some way of listening in on the last-minute planning sessions at Unscom's Baghdad headquarters in the Canal Hotel or there was a mole inside the organization. Hussein Kamel had unmasked Ekeus's translator as an Iraqi agent, but that individual had never had access to information as sensitive as this.

The Unscom offices in the Canal had been modernized in 1994 and were equipped with the best in countersurveillance technology that American and British intelligence could provide, making it unlikely that the Iraqis had succeeded in planting a bug. There was, however, a Russian scientist assigned to the teams who always seemed suspiciously inquisitive about upcoming "no-notice" inspections. Therefore, in strictest secrecy, a few of the senior members of the special commission staff planned and executed a sting operation. With only the suspect present, they discussed a purported upcoming surprise inspection at a specific location. Sure enough, discreet observation at the nominated site revealed the guards fully prepared for an Unscom visit. The Russian, who appeared to have been operating under the auspices of his country's foreign intelligence service, the SVR, was sent home amid conditions of deepest secrecy. The penetration of Unscom, using corruptible foreign intelligence agencies that Saddam had discussed years before with Wafiq al-Samarrai, had been brought to fruition, at least for a while.

If Iraqi intelligence had scored a coup against Unscom, the inspection agency had itself turned into a formidable intelligence organization. Things had come a long way since Rolf Ekeus had been forced to give a personal guarantee for the cash advance from the UN secretary general's special fund that had launched the organization.

"We became extremely successful at penetrating the concealment mechanism," says one former Unscom official. "We had gotten into [i.e., developed the ability to intercept] their communications. So we were only missing them by minutes. The Iraqis may have thought that the first time it might have been just luck, but the second, third, and fourth time it was obviously something more.

"This wasn't about biological weapons hidden in Saddam's palaces, as the press was suggesting. This was about the trucks moving around that moved the things we were after. When we went into places like the Special Republican Guard installations, we checked the drivers' logs to see who had been driving what truck and where. Our people knew almost by heart the names of the various people, drivers and so on, who were involved."

The units involved in the concealment effort did not operate in isolation. As Rolf Ekeus, the man the Iraqi leader once referred to as the "miserable spy," said after he left Unscom in July 1997: "It is the Special Republican Guards we are interested in, the concealment force. But they are also the protection force for Saddam. He can build new palaces, he can rebuild the weapons program, but he cannot replace the Special Guards, because they are the key loyal force. He does not have a replacement."

When the inspectors did manage to penetrate the compounds of this and other equally important units in pursuit of trucking records and other information germane to their enquiries, they inevitably saw evidence of other tasks assigned to these loyal servants of Saddam: lists of people to be arrested, logs of drivers who had transported prisoners to the grim confines of Abu Ghraib prison, duty rosters for standing guard at the palace. An inspector once opened a door in one of these complexes only to find a roomful of people sitting at desks wearing headphones. They were the telephone eavesdroppers. The inspector excused himself and closed the door.

Sometimes the interaction between Unscom officials and the Iraqi high command bordered on the surreal. Charles Duelfer, the Unscom second-in-command, recalls one occasion when he and Roger Hill, who succeeded Nikita Smidovich as chief of the concealment team, were making an exploratory survey of a presidential site. They had a map but were finding it difficult to figure out the perimeters of the site. Suddenly a large black Mercedes purred to a halt beside them. A rear window slid down, to reveal Abed Hamoud, the much-feared presidential secretary. Extending a genial greeting to the inspectors, Saddam's right-hand man asked if he could help. This was not as surprising as it may seem. Despite TV images of Unscom personnel engaged in grim, snarling face-offs with Saddam's minions, the two sides spent so much time in each other's company that they had inevitably become, if not friends, at least amiably civil with each other. Duelfer had even managed to strike up an amicable relationship with the previously shadowy Mahmoud, so he explained their problems with the map. "Let me see if I can help," said the Iraqi. Removing a large cigar from his mouth, he peered at the map and supplied helpful directions. Then, in response to a rapid command, one of his bodyguards opened the trunk of the limousine and reached inside. Duelfer wondered what the Iraqi leadership kept in the trunks of their cars—Kalashnikovs? rocket launchers?—and craned his neck to see. The bodyguard emerged with a tray of chilled Pepsis. After finishing their sodas, the Americans thanked their high-powered guide and he drove off.

Minutes later, another large Mercedes purred to a halt at the same spot. This one was white. The window slid down to reveal General Amer Rashid, minister for oil and the official responsible for negotiating with (and frustrating) Unscom. He too inquired as to what the two inspectors were doing. They explained that Abed Mahmoud had been most helpful in interpreting the map. "Nonsense," snorted Rashid. "Abed can't read a map. He probably had it upside down. Let me look." Duelfer explained that they were still puzzled by the precise placement of a particular boundary line. "Let's go see," said Rashid.

The line in question turned out to run along a thirteen-foot-high wall with a deep, square pit just in front. The Americans and Rashid surveyed it to check that they were in the correct place, all three stepping carefully around the pit. No one brought up the fact that the wall was heavily pitted with bullet holes, most of them grouped at chest height. Clearly, this was where the firing squads did their work, the pit providing temporary storage for dead bodies between shifts.

Any reference to the wall's gruesome function would have been "an intelligence question," i.e., raising a matter that lay outside Unscom's mandate and expressly barred from discussion. So, with the surveying completed, the two inspectors thanked Rashid and went on their way.

Rolf Ekeus finally left the organization he had created at the end of 1997. Richard Butler, the Australian diplomat who replaced Ekeus as special commissioner, promoted Scott Ritter to run the concealment inspections. The former marine was determined on an aggressive approach. He later described the Iraqi system of shifting the weapons and related materials from site to site, one jump ahead of the inspectors, as a "shell game." He declared that Unscom, rather than going after the shells (weapons, documents, etc.), should pursue "the man moving the shells." Not all the inspectors agreed with this single-minded tactic. "Ritter was obsessed with this notion that he was finally going to find the document that exposed 'the architecture of conceal-ment,'" says one official in close touch with the inspection effort. "But other people wanted to find out what the Iraqis were actually doing, and that meant looking for the weapons themselves."

By the summer of 1997, the main effort of these others was con-centrated on tracking the elusive remnants of the Iraqi biological and VX nerve gas programs effort as well as the possibility that Sad-dam might still have missiles and warheads with which to deliver these potent agents. Central to their concerns was the lack of evi-dence to support Baghdad's claim that it had destroyed the 25 mis-sile warheads and as many as 150 bombs it had filled with anthrax and botulinum toxin before the Gulf War. In addition, after sifting through the sites where Iraq insisted all its forbidden missiles had been secretly held in 1991 and 1992, Unscom announced that two remained unaccounted for.

Other bones of contention included seventeen tons of the "growth media" necessary for reproducing the toxins, nine hundred pounds of anthrax, and the possible existence of sprayers suitable for the technically highly difficult task of distributing the anthrax in fine enough particles to be absorbed in victims' lungs, as well as the true documentary history of the entire project. When Iraq submitted a sixth "Full, Final, and Complete Declaration" on the biological pro-gram in September 1997, Special Commissioner Richard Butler described it as "not even remotely credible."

The pressure from Unscom was matched by an increasingly defiant attitude from Iraq. In June 1997, an Iraqi "minder" accompanying inspectors in one of the Unscom helicopters seized the controls of the machine in order to prevent them from taking pictures of vehicles leaving a suspect site, causing a near crash. In the same week, another team was blocked from entering a site on instructions, said the Iraqi officials at the gate, "from the highest authority."

Following yet another censorious Security Council resolution demanding that Iraq cooperate with the inspectors, Saddam, sitting in council with the uniformed notables of the Revolutionary Command Council, issued a stern statement: "We would like to summarize and clarify our position as follows: Iraq has complied with and implemented all relevant resolutions. . . . There is absolutely nothing else. We demand with unequivocal clarity that the Security Council fulfill its commitments toward Iraq. . . . The practical expression of this is to respect Iraq's sovereignty and to fully and totally lift the blockade imposed on Iraq."

With hindsight, it is clear that the Iraqi leader had resolved to go on the offensive. All he needed was an excuse. That was to come soon enough.

In September, the obstruction of the inspectors grew more blatant. There was another helicopter incident on the thirteenth. On the seventeenth, a team hunting for details of VX production was kept outside the gate of the Iraqi Chemical Corps headquarters for hours while files were openly trucked away and other documents burned on the roof of the building. A week later, inspectors making a routine visit to a food-testing laboratory encountered several men carrying briefcases and trying to escape through a back door. Diane Seaman, an American microbiologist who was leading the team, seized one of the briefcases and opened it. Inside were kits for testing three deadly organisms as well as a logbook indicating that the lab had been conducting tests in secret for eight months under the supervision of the Special Security Organization.

By the end of October, the crisis was reaching a head. Insisting that Unscom had become no more than an espionage agency operating on behalf of the United States to prolong sanctions, Tariq Aziz announced on October 29 that no more Americans would be allowed into Iraq to work on the inspection teams. Four days later, he announced that the

U-2 high-altitude photo-reconnaissance plane lent by the United States to Unscom was operating as a spy plane for the Americans. (Aziz was presumably unaware that by this time Scott Ritter was routinely—with the approval of his superiors—sharing U-2 photo intelligence with the Israelis.) These threatening statements were followed by news that the Iraqis were sabotaging the work of the Unscom long-term monitoring effort, in which erstwhile weapons sites were surveyed by remote cameras to ensure that forbidden work had not resumed. A few days later, the remaining American inspectors were expelled from Iraq.

As a U.S. military riposte to this defiance appeared to be increasingly inevitable, the familiar features of an Iraqi crisis reappeared. Once again, Saddam's picture adorned the covers of news magazines. *Time* declared somberly that this was "the gravest international crisis of [Bill Clinton's] presidency." On television and in print, biological warfare experts solemnly described the massacres that could be perpetrated with only a minute fraction of the Iraqi leader's presumed stockpile of anthrax. Eminent columnists began sounding like Tikritis, calling glibly for a "head shot" against Saddam, while the nightly network news displayed stirring scenes of the U.S. military gearing up for action. The atmosphere summoned up memories of the Gulf War, when White House correspondents asked President Bush in all seriousness why he was not making greater efforts to kill the president of Iraq.

The reality, however, was very different from those heady days. Most important, the coalition built by George Bush had almost completely disappeared. This time the Saudis made it clear that they did not even want to be asked to let their territory be used by U.S. warplanes in bombing Iraq. The United States did not dare ask the Security Council for the authorization to launch an attack for fear that Russia or France, both increasingly sympathetic to Iraq's position, would cast a veto. As it was, Washington was "stunned" by the indifference of the Security Council to Saddam's expulsion of the American inspectors. The most severe sanction the council was willing to pass was a ban on international travel by Iraqi weapons scientists, the last people Saddam was likely to allow to leave the country.

The Clinton administration insisted that it had every right to bomb Iraq under existing resolutions, and prepared targeting plans. But here again, President Clinton and his advisers were faced with problems. The targets attacked in the first days of the Gulf War had been easy to choose—power plants, the presumed centers of Iraqi nuclear and other mass-destruction weapons production, Saddam Hussein himself. Subsequent inquiries had revealed that while the power-plant bombings had done permanent damage to Iraq's civilian infrastructure, they had not brought down the regime or even hindered its military capabilities to any great extent. The most important weapons plants—al-Atheer for nuclear, al-Hakam for biological production—had not even been targeted, let alone destroyed. Saddam and all other important officials had simply stayed away from obvious targets and had escaped unscathed. The rationale advanced by the White House for attacking this time was that if Saddam was preventing Unscom from rooting out his weapons capabilities, the job would have to be done with high explosives. But no one knew where these weapons and systems were actually hidden at any particular time. Some of the suspected production facilities were "dual use," with legitimate civilian applications in, among other places, hospitals. The United States could hardly bomb them.

As Clinton and his advisers mulled over these awkward choices, Saddam chose to back off, at least for the moment. Having tested the strength of the U.S. alliance, he chose to accept mediation from an old friend, Russian foreign minister Yevgeny Primakov. Primakov pledged to press for the lifting of sanctions. In return, the Iraqi leader agreed that American inspectors could return to Baghdad. The Clinton administration greeted the news with relief. By November 20, the immediate crisis was over.

From Saddam's point of view, the confrontation had yielded eminently satisfactory results. The United States had declared that Unscom's right to inspect was an issue on which it was prepared to go to war, and had then found itself, except for the British, entirely bereft of useful allies. Unscom, from being a threat to the Iraqi leader, had turned into an advantage. He now had the initiative because he could provoke a confrontation any time he chose, simply by refusing to cooperate. In pursuing this strategy, Saddam had an unlikely ally (albeit one with a different agenda), Scott Ritter, who returned to Baghdad on November 21 as determined as ever to search for Saddam's secrets, regardless of the consequences—which as likely as not would be renewed Iraqi obstruction that in turn would necessitate a forceful U.S. response.

The realization that the power to provoke crises appeared to rest in the hands of President Saddam Hussein and Major Scott Ritter had by this time dawned on the Clinton administration. An aggressive effort by Ritter, just when they had drawn back from military action in the hope of garnering more international support, was not at all what was required. According to Ritter, Special Commissioner Richard Butler now came under heavy pressure to rein in the energetic inspector, pressure to which he yielded. State Department and Unscom officials indignantly deny that Butler was following instructions from outside. "It wasn't just Madeleine Albright who didn't want Scott Ritter starting a crisis whenever he felt like it," says one State Department official indignantly. "Richard Butler didn't want him doing that either." For whatever reason, for the time being Butler canceled Ritter's planned inspections.

Meanwhile, both the United States and Iraq were readying themselves for a fresh confrontation. The Clinton administration had concluded that Saddam had gotten the better of the United States in the November crisis and the Pentagon was dusting of its target lists. The casus belli would be the principle of access for Unscom to the eight sprawling complexes comprising Saddam's somewhat gaudy palaces, security forces offices, and barracks, as well as other government facilities generically referred to as "presidential sites." When Iraq announced it was denying access to these areas, the United States rose to the bait and embraced the sites as the defining issue.

Saddam was quite ready for a second round. In November, the government had admitted the foreign press en masse, a move that yielded ample dividends in the form of sympathetic descriptions of the plight of the Iraqi people after seven years of sanctions. Now Baghdad began to fill up once again with journalists and TV crews from around the world. By mid-February, the number had reached eight hundred. Their all-too-accurate depiction of hospitals without medicines, schools without books, and mothers without food for their children had a searing impact on international public opinion. Pope John Paul II eloquently expressed the feelings of many when he declared, in an address to the Vatican diplomatic corps in January 1998:

"As we prepare for a new round of bombings, we cry out in anguish over seven years of United Nations sanctions against the Iraqi people, which can only be understood as biological warfare against a civilian population. During the Gulf War, U.S.–led coalition forces deliberately targeted Iraq's infrastructure, destroying its ability to provide food, water and sanitation to its civilian population and unleashing disease and starvation on an unimaginable scale. United Nations reports claim that over 1 million civilians have died as a direct result of the sanctions. UNICEF reports that 4,500 children are dying each month. As people of faith, we are ashamed that the actions of the UN, whose mission is to foster peace, can be so deliberately directed toward the sustained slaughter of innocent civilians."

Nevertheless, U.S. officials pressed on doggedly to make the case and gain the necessary support for bombing. Defense Secretary William Cohen sought to rouse European officials by stating that "a poison that kills" (actually ricin, the most toxic substance known) can be extracted from "six or seven castor beans," also the source of castor oil. In Iraq, Cohen noted darkly, "they are growing hundreds of acres of castor beans," leaving his audience wondering if bean fields were being added to the target list. A stream of high-ranking American officials touring the capitals of the Gulf states failed in many cases to extract even the mildest endorsements for an American attack. The most telling rejection came from Bahrain, the tiny island-state lying off the coast of Saudi Arabia, long a staunch American ally and the staging area for Unscom since its inception. President Clinton had personally spoken with the emir to ensure his support. Even so, the Bahraini information minister issued a statement declaring that the United States could not attack Iraq from his country.

The Arab leaders had not come to love Saddam in the seven years since the Gulf War. Their chilly attitude toward the American pleadings was derived from the fact that no U.S. strike was likely to get rid of the Iraqi leader and also the growing public outrage among their subjects over the suffering of ordinary Iraqis. In 1990 and 1991, the public in the Gulf and the rest of the Arab world had been comparatively deprived of access to information. (The Saudi government withheld news of the invasion of Kuwait from its citizens for forty-eight hours.) They could listen to the BBC or Radio Monte Carlo for news that their rulers preferred to keep out of the local (tightly controlled) media, but such an audience was, in most cases, limited. In the 1990s, however, the region had been swept by a communications revolution. Arab-language satellite TV channels brought comparatively uncensored news into the homes of anyone with a dish. The uncontrollable Internet served the same function. The public, thus informed, was resolutely against support for the United States and its perceived agent, Richard Butler, in raining more bombs on Iraqi children already decimated by sanctions. Even the most absolute of monarchs had to pay attention.

The effect of changing patterns in communications was further brought home to the administration when Madeleine Albright, William Cohen, and National Security Adviser Sandy Berger attempted to market their policy at a "town hall meeting" at Ohio State University. The event was a fiasco. Amid continuous heckling, angry citizens challenged America's "moral right" to bomb Iraq. The proceedings, humiliating for Albright, Cohen, and Berger, were televised internationally by CNN. Iraqi TV ran them in full.

By coupling the issue of a putative secret Iraqi missile force armed with biological weapons with the issue of the presidential sites, Washington had given a hostage to fortune. "All we ever believed was in these places," says Unscom deputy chief Charlie Duelfer, "was documents." Any weapons were almost certainly hidden elsewhere. But the impression took hold among press, public, and politicians that Saddam was concealing the deadly missiles in the recesses of his infamous palaces, immune from the attentions of the inspectors. In an ill-advised remark to the *New York Times*, Richard Butler suggested that such missiles could be fired "at Tel Aviv," thereby igniting panic in the Israeli capital, where long lines quickly formed to pick up gas masks and the government rushed in 6 million doses of anthrax vaccine from the United States. But the United States itself did not really appear to take the threat of an Iraqi biological or chemical missile strike very seriously. In Kuwait, which would presumably have been high on Saddam's list of possible targets, U.S. citizens were advised by their embassies that there was no cause for alarm and certainly no need to equip themselves with gas masks.

In the months between the invasion of Kuwait and the outbreak of the Gulf War, the Bush White House had been haunted by the fear of a "diplomatic solution" that would allow Saddam to extricate himself from Kuwait without undue loss of face. In those days, the United States, aided by the Iraqi leader's intransigence, had ruthlessly quashed any initiatives aimed at such a solution. But by February 1998, the world had changed. The French, who in any case had been busily negotiating business relationships with Iraq, had argued that there was little point in an inconclusive military action that would not get rid of Saddam. Now they suggested that UN secretary general Kofi Annan travel personally to Iraq to seek a way out of the crisis over the presidential sites.

Annan thought this an excellent idea. Washington did not. "You can't go," UN ambassador Bill Richardson told Annan. "It will box us in." But even the British thought the secretary general should be allowed to go to Baghdad. Clinton agreed with reluctance.

The secretary general's trip was a breakthrough for Saddam. For the first time since the war, a world statesman was coming to visit, addressing him respectfully and seeking a favor. The Iraqi leader speedily agreed to a compromise under which Unscom could inspect the presidential sites, but only when accompanied by a newly formed team of diplomats who would monitor the activities of the obstreperous inspectors. Thus, rather than asserting the principle of free and unfettered access to any site that Unscom needed, the agreement created a new and cumbersome procedure for this special category of site.

None of this mattered to Annan. After he had smoked one of Saddam's cigars, the secretary general described his host as "calm" and "very well informed and . . . in full control of the facts."

Since the crushing of the 1991 uprisings, Saddam had rarely been seen in public. Now, in the fullness of his triumph, he embarked on a program of public appearances. On March 17, for example, he visited al-Dhour, a small town in the Sunni heartland. This locality held a special significance in the Saddam Hussein story because it was here, in 1959, that he had swum the Tigris following his abortive attempt to assassinate President Qassim. He took phone calls from local citizens and accepted the "greetings of the masses, who received him with shouts of praise and dancing," according to

Iraqi TV. Afterward the crowd slaughtered sheep in celebration while the leader waved from the back of an open car, firing his rifle in the air again and again.

The fact that Annan's visit had endowed Saddam with a legitimacy he had not enjoyed in years was not lost on the Republican Party leadership in Washington. Denouncing the administration's weak acquiescence to Annan's "appeasement," the Republicans in Congress looked for a means to discommode both Clinton and Saddam simultaneously and found it in none other than Ahmad Chalabi.

Ever since the CIA had withdrawn funding from the Iraqi National Congress at the beginning of 1997, the opposition group had fallen on hard times. Chalabi claimed to be supporting the opposition group out of his own pocket, to the tune of no less than $5 million, but the INC London headquarters had taken on a semi-deserted look. The once-bustling INC center at Salahudin in Kurdistan had been abandoned since the massacre and headlong flight of September 1996. As an active opposition movement within Iraq, the INC was defunct. Nevertheless, to powerful senators like Trent Lott and Jesse Helms and their advisers, including the formidable cold-war veteran Richard Perle, the articulate Chalabi was a godsend.

Speaking as an "elected representative" of the Iraqi people (a claim based on a vote by the three hundred delegates at the inaugural INC meeting in Salahudin back in October 1992), Chalabi told a Senate committee that the INC was "confronting Saddam on the ground" and had the support of "thousands of Iraqis." After leveling some abuse at the CIA and paying tribute to the "warrior" Scott Ritter, he proposed that the United States should deploy its forces to establish "military exclusion zones" in northern and southern Iraq. The northern zone he had in mind was far larger than the area controlled by the Kurds, including the major cities of Mosul and Kirkuk and Iraq's northern oil fields. The southern area included Basra and the southern oil fields. The INC would take over the administration of these areas, assisted by the United States, and would eventually establish itself as the provisional government of Iraq. The whole undertaking would be financed either from Iraqi assets frozen in U.S. banks since 1991 or by the sale of oil from the southern zone.

This ambitious scheme went down well with the Senate majority party. A Democrat who had the bad taste to bring up the issue of the embezzlement charges against Chalabi in Jordan following the collapse of the Petra Bank in 1989 was roundly abused by the former banker's supporters, along with a suggestion that even the mention of this event "had the earmarks of a plant from the White House or the CIA." In the following months, support for the Iraqi opposition and Chalabi in particular blossomed in Congress, which voted $5 million to establish a "Radio Free Iraq" along the lines of the Radio Free Europe that had been beamed into Eastern Europe during the cold war. Another $5 million was voted for the "Iraqi democratic opposition," with the proviso "that a significant portion of the support for the democratic opposition should go to the Iraqi National Congress, a group that has demonstrated the capacity to effectively challenge the Saddam Hussein regime with representation from Sunni, Shia, and Kurdish elements of Iraq."

Thus, while many of its former leading Iraqi members—including the Kurdish leaders Barzani and Talabani—considered Chalabi's organization extinct, the INC, as a weapon in the Republicans' armory, was going from strength to strength on Capitol Hill. For the first time since the debate that preceded the Gulf War, Iraq had become a partisan issue in U.S. politics.

This being the case, the administration had to fight back. Officials briefed journalists on the all too evident weakness of the opposition, including the INC. Others leaked word that the CIA was hard at work on a whole new covert scheme of "sabotage and subversion" to undermine Saddam. Kurdish and Shiite agents would be enlisted to destroy "key Iraqi pillars of economic and political power, like utility plants or broadcast stations." Whoever was responsible for this "plan" had evidently forgotten Abu Amneh, the mercenary bomber and self-proclaimed veteran of the last CIA covert action against Saddam. Nor did the mooted scheme indicate much knowledge of contemporary conditions inside Iraq, where the utility plants were failing without the need of any outside intervention by the CIA. By the summer of 1998, the hottest in fifty years in Iraq, even the power plants in Baghdad were regularly out of action for twelve hours and more a day.

On a more practical level, the administration made efforts to reach out to Saddam's enemy to the east, Iran. For years, the "dual containment" policy, by which the Iranians were accorded equal status as pariahs with the Iraqis, had precluded any effective collaboration between Tehran and Washington. But by 1998, the cold war between Washington and Tehran showed signs of winding down, aided by appeals for better relations from the liberal cleric Mohammed Khatami, elected president of Iran in May 1997. Accordingly, Mohammed Baqir al-Hakim, the leader of the Iranian-backed Supreme Council for the Islamic Revolution in Iraq, began to receive earnest appeals to visit Washington. Hakim rejected these overtures, presumably with the encouragement of the powers that be in Tehran. The Iranian authorities were not about to help solve Washington's Iraq problem without receiving something tangible in return, such as U.S. blessing for shipment of Central Asian oil across Iranian territory.

At the same time, the State Department moved to rebuild old alliances. Before August 1996, northern Iraq had been a "military exclusion zone" denied to Iraqi forces. Ruminating on various possible means of challenging Saddam, the State Department now took steps to restore the status quo in Kurdistan. Accordingly, in early September 1998, Massoud Barzani and Jalal Talabani were invited to Washington for a peace meeting, lodging at the Key Bridge Marriott Hotel. In return for a firm guarantee of American military protection against Saddam, the two leaders agreed to swallow their mutual enmity once again and unite in a reformed Kurdish government, with elections to follow. Barzani agreed to share the money from the border-crossing tolls and Talabani agreed that Arbil should be jointly controlled by the two groups. As Barzani and his delegation came and went through the Marriott lobby during the negotiations, they were surveyed with vocal enmity by Jahwar al-Sourchi who, by coincidence, had booked himself into the same hotel while in Washington on a business trip. As he muttered imprecations against his tribal enemies, the blood feuds of the distant mountains of Kurdistan seemed suddenly very near.

Chalabi greeted the initial news of the Kurdish agreement with exultation. "Things are really moving," he said the day after the agreement was announced. But these high hopes were dashed when the Kurdish leaders flatly refused to have anything to do with him. Even an imperious summons from the office of Senator Jesse Helms for the pair to come to a joint meeting with Chalabi could not sway them. The discussion with Helms's messenger became acrimonious, with ugly words such as "embezzler" being tossed about. It did not appear that the INC would be returning to Salahudin anytime soon. To add to Chalabi's vexation, the CIA leaked word that the agency's inspector general was investigating the agency's prior handling of both the INC and the Accord operations, including the use of funds. This did not sway Chalabi's partisans in Congress, however, who by October had passed the "Iraq Liberation Act," authorizing $97 million for the arming and training of the Iraqi opposition. Precisely where this training was to take place and who would be trained was not specified.

Meanwhile, reviewing the recent crises over Unscom, U.S. officials concluded that the confrontations with Saddam had been a disaster. In late April, President Clinton secretly decreed that, for the time being, there would be no more attempts at military action to force the Iraqis to allow access to presidential sites or anywhere else to the Unscom inspectors. Even when tests on a missile warhead excavated from one of the sites where Iraq had secretly destroyed weapons in 1991 indicated that it had once contained VX, thus giving the lie to Iraqi denials that it had ever succeeded in "weaponizing" the lethal chemical, the administration showed little appetite for an immediate face-off.

In Baghdad, Saddam was stepping up the level of his rhetoric by demanding a speedy conclusion of the Unscom mission and threatening grave but unspecified retaliation if sanctions were not lifted. Unscom was still going about its work, seeking elusive documents and other evidence of Iraqi perfidy. On August 5, however, the Iraqi government announced that it was ending all cooperation with the inspectors, thus ending their searches. The White House, true to the April decision to swear off military confrontations over the issue, had little reaction.

By now, Washington knew for certain that Saddam had been deliberately seeking a provocation. An electronic intelligence interception of a conversation between Tariq Aziz and Russian foreign minister Primakov revealed Aziz angrily complaining that "the Americans are not reacting" to the action against the inspectors. If the fact that the recent intrusive searches had been to the advantage of Saddam was now clear to high-level officials such as Madeleine Albright, the point was irrelevant to Scott Ritter. On August 27, he resigned, citing the interference with his work by high-level officials in Washington and London and complaining that "the illusion of arms control is more dangerous than no arms control at all." With this and subsequent denunciations of the administration's weakness in the face of Saddam's defiance, the articulate ex-marine swiftly became as much a hero as Ahmad Chalabi to the Republicans, anxious as they were to malign Clinton administration policy on Iraq.

Ritter lost no time in asserting the magnitude and imminence of the threat from Saddam's hidden arsenal, declaring that Saddam had at least three nuclear weapons ready for use as soon as he laid his hands on the necessary fissile material (uranium 235 or plutonium). This was too much for many of his former colleagues on the inspection teams. Gary Dillon, leader of the "action team" deployed by the International Atomic Energy Agency (the IAEA) to work on the specifically nuclear aspects of Iraq's weapons programs, asked Ritter how he had learned of these three nuclear devices. "From a northern European intelligence source," replied Ritter. The response from the nuclear experts was laughter.

"For political reasons, the United States pushes the IAEA to find little discrepancies in Iraq's nuclear accounting so that the file can be kept open," explains one official closely involved in the operation, "but short of lobotomizing or killing all the Iraqi nuclear scientists, the Iraqi nuclear program is finished. We have closed down all their nuclear facilities and activities."

Having achieved fame as the Unscom martyr, Ritter now inflicted another wound on the organization. In an interview with the Israeli newspaper Haaretz, he spoke in glowing terms of his close and fruitful relationship with Israeli intelligence, as well as detailing such hitherto closely held Unscom secrets as the organization's ability to monitor Iraqi communications. On the same day his Haaretz interview appeared, the Washington Post reported that Ritter, with his superiors' approval, had been in the habit of bringing film taken by Unscom's U-2 spy plane to Israel for processing and analysis. Only a few months before, the United States had been seeking to rally Arab support in asserting Unscom's right to inspect at will. Given this admission of collusion with Israel, however well intentioned, the prospect of any Arab support for Unscom were clearly fading away.

As they maneuvered, both sides were using Unscom as a tool. On November 1, Saddam upped the ante by suspending all cooperation with Unscom's long-term monitoring program, meaning that the inspectors could no longer check to ensure that sites already visited were not being used for work on weapons. Events now followed a familiar pattern. The United States and Great Britain announced that they were ready to bomb Iraq. Statements of defiance poured forth from Baghdad. At the very last minute, with U.S. warplanes actually in the air on their way to attack Iraq, the Iraqi government offered to resume "full co-operation" with Unscom. The bombers returned to their bases, but only for a brief period.

The Clinton Administration and Saddam Hussein, it appeared, were both intent on fomenting the much postponed bombing attack. Richard Butler's inspectors returned to Baghdad and went about their searches. Most of these passed off without incident, but the Iraqis on some occasions provided just enough non-cooperation to justify Butler's subsequent report that Saddam had once again failed to live up to his commitments. Reliable reports at the time suggested that Butler had composed his report in close consultation with Washington. Indeed the vociferous Scott Ritter went on record with the claim that the inspections had been a "set-up," designed to "generate a conflict that would justify a bombing." Saddam for his part, in insisting that Butler stick to the letter of the agreement negotiated by Rolf Ekeus in June 1996 and send no more than four inspectors to sensitive sites such as Baath Party headquarters, appears to have been no less eager to have the bombs fall.

In Washington, of course, everything was overshadowed by the ongoing impeachment proceedings against President Clinton. When he duly ordered the long heralded bombing strike on December 16, his Republican opponents reacted with angry suspicion, claiming with some justice that the attack had been timed to serve as a distraction from the president's problems at home. However, apart from the postponement of the House of Representatives' debate on impeachment by one day, the attack on Iraq was of little political benefit to the commander in chief.

The bombing elicited furious protests from France, Russia, China, and Egypt, while angry crowds demonstrated in the Arab world on behalf of the Iraqi people. Palestinians set fire to the American flags they had been given to wave in honor of President Clinton's visit to Gaza only a few days before. Nor was the attack effective in humbling Saddam or eliminating his alleged arsenal of weapons of mass destruction. Ninety seven targets overall were attacked, of which only nine were reported by the Pentagon as fully destroyed. Of eleven chemical and biological weapons production facilities targeted, none were destroyed. The Special Republican Guards and other bastions of the regime associated with weapons concealment were similarly slated for destruction, but even assuming they had not evacuated their peacetime barracks and offices as they did in January 1991, the results in terms of facilities destroyed appear to have been meager.

The Pentagon expressed surprise at the lack of antiaircraft fire, but Iraq's most effective defenses were the massed ranks of television news cameras from around the world on the roof of the international press center in Baghdad. Under such scrutiny, the United States could not risk high-profile "collateral damage" such as the attack on the Amariya shelter that had incinerated four hundred women and children eight years before. In Baghdad itself, people greeted the renewed offensive with weary resignation. "Iraqis," as one of them remarked, "fear that a game is being played over which they have no interest. They feel they are always the victims, whether it is sanctions or bombs." The streets emptied as the air-raid sirens wailed at nightfall, but wedding parties continued at the al-Rashid hotel and the Iraqi dinar in contrast to previous crises, retained its value against the dollar. "Operation Desert Fox," repeatedly threatened and postponed for more than a year, had turned out be only a shabby and diminished echo of the storm unleashed on Iraq in the distant days of January 1991.

Dr. Hussain al-Shahristani, the man who had defied Saddam's orders to build a nuclear weapon so many years before, was living in Tehran. His dedicated work on behalf of Iraqi refugees endowed him with considerable moral authority among the Iraqi Shia and a wide range of contacts inside Iraq, especially in the south. Two days into the bombing he wrote one of the present authors an urgent message. "A number of people have contacted us from inside Iraq," he wrote, "and asked if the Americans are really going to continue this [bombing] campaign to weaken Saddam to a point where people can rise up and free themselves from the regime. The memory of betrayal during the last intifada is vivid in people's minds, and they do not want to repeat that tragic experience." The Iranian government had made its own attitude clear by closing its border with Iraq in order to prevent any assistance to a potential uprising.

Following seventy hours of bombing, President Clinton gave al-Shahristani and his people their answer. He declared victory—"I am confident we have achieved our mission"—and called off the attacks. Saddam Hussein also pronounced himself the winner. "God rewarded you," the Iraqi leader told his subjects in a TV address that was broadcast across the Arab world, "and delighted your hearts with the crown of victory." Iraqi spokesmen insisted that there would be no further cooperation with Unscom.

"Operation Desert Fox," as the bombing offensive was officially named by the Pentagon, marked the end of an era in which the presence and activities of Unscom in Baghdad permitted Saddam to create crises with the United States and inject himself into the global headlines whenever he wished.

Thereafter, Washington decided that leaving Saddam out of sight and therefore out of the minds of the public was the least undesirable option. The new policy was a resounding success. Though the U.S. continued to bomb Iraq, the raids were for the most part in regions distant from Baghdad and officially billed as defensive measures in response to Iraqi "provocations" in threatening U.S. and British warplanes. After a short period, news reports of these routine raids were relegated to a few brief lines deep inside the newspapers.

In February, 1999, Saddam gave proof, if proof were needed, of his absolute control of non-Kurdish Iraq. Ayatollah Mohammed Sadeq al-Sadr, a Shi'ite teacher from Najaf, had originally been fostered by the regime as a counterweight to the influence of the Shia religious establishment in the holy cities. But al-Sadr's influence had grown by leaps and bounds, to the point where he had become an independent voice of real influence. In his weekly sermons,

echoed across southern Iraq and even in the Shia suburbs of Baghdad, he criticized the government, as well as the Americans and
Arab states, for deepening the miseries of his flock. Although he
had refrained from criticizing Saddam directly (Uday was not so
lucky) the end result was inevitable. On February 10, 1999, he and
his two sons were ambushed and shot dead in their car on the outskirts of Najaf.

No one doubted that Saddam had ordered the killings (despite
denials from Baghdad) and for a few days the south of Iraq erupted
in scenes of fury reminiscent of March, 1991. There were riots in
Baghdad. Artillery fire echoed in the suburbs of Basra. Rumors circulated on the borders that rebels had seized Nassariyah. For a
brief moment it appeared that the frustration and rage of ordinary
Iraqis had boiled over. But the inevitably savage crackdown soon
quelled the disturbances and the country relapsed into its former
state of stagnant calm.

Meanwhile in Washington, the definitive end of Unscom inspections was followed by a series of embarrassing revelations concerning the extent to which the supposedly independent UN agency
had been co-opted by the CIA for its own purposes. The remote
monitoring system erected by UNSCOM to watch former mass
destruction weapons centers, for example, turned out to have been
used to relay Iraqi military communications for the benefit of US
intelligence, thus vindicating all Saddam's accusations that the
inspectors were simply acting as American spies.

In declaring victory at the end of "Operation Desert Fox," President Clinton promised to "sustain what have been among the most
intensive sanctions in UN history." It was an ominous warning, fully
justified by the continuing purgatory of the Iraqi civilian population. Amid the furor over Scott Ritter's resignation in the summer
of 1998, another resignation passed with little attention. Denis Halliday, the Irish Quaker who had been sent to Baghdad to monitor
the oil-for-food arrangement under which revenues from exports of
Iraqi oil were entrusted to the custody of the United Nations to buy
food and other humanitarian supplies, was leaving Baghdad in dis-
gust. As he left, he directed a bitter blast at the policy that caused
"four thousand to five thousand children to die unnecessarily every

month due to the impact of sanctions because of the breakdown of water and sanitation, inadequate diet, and the bad internal health situation."

In her March 1997 speech at Georgetown announcing the indefinite continuation of sanctions, Secretary of State Albright had described the oil-for-food deal just then coming into effect as being "designed to ease the suffering of civilians throughout Iraq." As it so happened, the month after she spoke, UNICEF conducted a survey of some fifteen thousand children under five across Iraq. The results showed little difference between the cities and the countryside. Just under a quarter of the children were underweight for their age. Slightly more than a quarter were chronically malnourished. Almost one in ten was acutely malnourished. In March 1998, after the oil-for-food program had been in effect for twelve months and indeed had been vastly increased in value, UNICEF did another similar survey. The percentage of underweight children had gone down by a statistically insignificant margin. Those with chronic malnutrition had declined by eight tenths of a percentage point, while the acutely malnourished infants and toddlers had actually increased by a tiny fraction. Commenting on these chilling statistics, the authors of the report noted in bold type that "It would appear that the 'oil-for-food' program has not yet made a measurable difference to the young children of Iraq."

Though the horrifying conditions endured by ordinary Iraqis aroused little or no concern in the U.S., they did constitute an embarrassment for the proponents of sanctions, especially at the UN. In September 1999, in advance of what turned out to be a sterile debate at the Security Council, the State Department released "Saddam Hussein's Iraq," an effort to persuade the world that any and all human misery in Iraq was the sole fault and responsibility of the Iraqi leader. "The international community, not the regime of Saddam Hussein, is working to relieve the impact of sanctions on ordinary Iraqis," claimed the authors of the report in their preamble.

Key to this self-justification was the oil for food program, a mechanism understood by few apart from those directly concerned. Proceeds from the oil sales were banked in New York (at the

Banque National de Paris). Thirty-four percent of the money was set aside for disbursement to outside parties with claims on Iraq, such as the Kuwaitis, as well as to meet the costs of the UN effort in Iraq. A further thirteen percent went to meet the needs of the Kurdish autonomous area in the north. Iraqi government agencies, meanwhile, under consultation with the UN mission resident in Baghdad, drew up a list of items they wished to buy. This list could include food, medicine, medical equipment, infrastructure equipment to repair water and sanitation etc., as well as equipment for Iraq's oil industry. UN hq in New York reviewed the list, approving or disapproving specific items. Then the Iraqis ordered the desired goods from suppliers of their choice. Then came the most crucial step in the process. Once the Iraqis had actually placed an order, the contract went for review to the 661 Committee. This was made up of representatives of the fifteen members of the Security Council and is named for Security Council Resolution 661, which originally mandated the sanctions, on August 6, 1990. The committee had the power to approve or disapprove (although the preferred euphemism was to put "on hold") any of the contracts. Approved contracts were then filled by the supplier and shipped to Iraq, where they were inspected on arrival by an agency called Cotecna. When this agency certified the goods have arrived, the supplier is paid from the oil cash in the bank in New York.

"Since the start of the oil-for-food program," declared the State Department report, "78.1 percent [of the contracts submitted for review to the 661 Committee] have been approved." That meant that 21.9 percent of the contracts were denied, with the overwhelming majority of the vetoes imposed by the U.S. and Britain. The 448 contracts on hold as of August 1999, the State Department report explained, included items that could be used to make chemical, biological and nuclear weapons.

Anyone who understood the situation in Iraq knew that the root cause of child mortality and other health problems was no longer simply lack of food and medicine but the lack of clean water (freely available in all parts of the country prior to the Gulf War) and of electrical power, which as of September 1999 running at 30 percent of the pre-Gulf war level. Of the 21.9 percent of contracts vetoed by the 661 Committee, a high proportion were integral to the efforts to repair the water and sewage systems. The Iraqis had submitted contracts worth $236 million in this area, of which $54 millions, worth roughly one quarter of the total value, had been disapproved.

"Basically, anything with chemicals or even pumps is liable to get thrown out," one official from the UN's Office of the Iraq Program told the authors. The same trend was apparent in the power supply sector, where around 25 percent of the contracts were on hold—$138 million worth out of $589 million submitted. The proportions of approved/disapproved contracts did not tell the full story. UN officials referred delicately to the "complementarity issue," meaning that items approved for purchase may be useless without other items that had been disapproved. For example, the Iraqi Ministry of Health had ordered $25 millions worth of dentists' chairs, said order being approved by the 661 Committee—except for the compressors, without which the chairs were useless and consequently gathered dust in a Baghdad warehouse.

The State Department report made much of the vast quantities of medical supplies (including the dentist chairs) sitting in Baghdad warehouses, implying that Saddam was so cruelly indifferent to the suffering of his subjects that he preferred to let them die while stockpiled medicine went undistributed. "They don't have forklifts," countered one U.N. official involved with the program. "They don't have trucks, they don't have the computers for inventory control, they don't have communications. Medicines and other supplies are not efficiently ordered or distributed. They have dragged their feet on ordering nutritional supplements for mothers and infants, but it's not willful. There is bureaucratic inefficiency, but you have to remember that this is a country where the best and the brightest have been leaving for the past nine years. The civil servants that remain are earning between $2.50 and $10 a month." The breakdown of the Iraqi communications system—it was taking two days to get a phone call through to Basra from Baghdad—was obviously a fundamental impediment to the health system, but when the Iraqis ordered $90 million worth of telecommunications equipment, all of it was vetoed. The justification was that Saddam could use the system to order troops about, notwithstanding the fact that his security services had the use of their own cell-phone system, smuggled in from China.

For Denis Halliday, who continued to speak out against sanctions after he had resigned, the hungry and dying children in a land where overeating had been the major prewar pediatric problem were only the most obvious effect of the United Nations blockade. Sanctions, he said, were biting into the fabric of society in less visible but almost equally devastating ways. They had, for example, increased the number of divorces (up to 3 million Iraqi professionals had emigrated, leaving their womenfolk behind to head the household) and reduced the number of marriages because young people could not afford to marry. Crime had increased. An entire generation of young people had grown up in isolation from the outside world. He compared them, ominously, to the orphans of the Afghan war who had spawned the cruel and fanatical Taliban movement. These young Iraqis were intolerant of what they considered to be their leaders' excessive moderation. "What should be of concern is the possibility of more fundamentalist Islamic thinking developing," concluded Halliday. "It is not well understood as a possible spin-off of the sanctions regime. We are pushing people to take extreme positions."

Following the suppression of the great insurgencies of 1991, Saddam Hussein had announced his intention of sitting back and waiting to take advantage of his enemies' mistakes. In the ensuing years, those mistakes had been plentiful. Saddam himself had survived unscathed. But the biggest mistake of all was to make the Iraqi people pay the price of besieging Saddam.

One day, the bill will come due.

Postscript

The United States fought the Gulf War to prevent change. Saddam Hussein, for so long a useful regional ally, had upset the natural order of things by invading Kuwait on August 2, 1990, thereby threatening Western control of Middle Eastern oil reserves. And so, armies, fleets, and bombers were sent to turn the clock back—to August 1, 1990. Once this was accomplished, Iraq would resume its former role, albeit crippled by the conflict and shorn of its most dangerous weapons, but still united and potent enough to continue countering revolutionary Iran. Any changes in the government of Iraq, so Washington hoped, would be confined to the smooth removal of Saddam himself by means of a military coup. Pending a turnover of leadership, Saddam would be contained, as administration officials liked to say, "in his box." The Iraqi people were to be locked in with him, out of sight and out of mind, for the duration of his rule.

But the clock could not be turned back. Whatever the hopes of the victors, the war had brought inevitable and irreversible changes.

Western public opinion, suddenly educated as to the more monstrous aspects of Saddam's regime, would no longer permit the brutal repression of the Kurdish minority, thereby forcing the United States to sponsor a semiautonomous Kurdish statelet in northern Iraq.

The Iraqi opposition could no longer be spurned and derided in Washington and other allied capitals as before. They were therefore recruited as supporting players in the CIA's covert program to organize a coup. Weak inside Iraq itself, the exiled opposition groups yet served as an increasingly vocal and embarrassing reminder that American policy toward Iraq was having the effect of preserving the status quo.

Most important, Saddam himself consistently refused to play his assigned role. Although weakened by war and rebellions, he did not fall victim to a putsch by his Baathist peers. The resourceful dictator showed great skill in manipulating divisions among the Sunni, Shia, and Kurdish communities in Iraq to his own advantage. At his moment of greatest danger in March 1991, the Sunni officer corps in Baghdad demonstrated that they preferred Saddam to the rebels who had captured northern and southern Iraq. Nor was Saddam prepared to acquiesce in the decision to destroy his strategic nuclear, chemical, and biological weapons programs. Resisting with artful cunning, he retreated only when faced with the real threat of renewed allied military action, as in the summer of 1991, or betrayal from inside, as with the defection of Hussein Kamel four years later.

Even so, Saddam stayed on the defensive until 1996. Then he sent his tanks into the Kurdish capital of Arbil and proved to his satisfaction that the United States was unwilling to intervene. From that moment on he felt free to provoke repeated confrontations with his enemies. In December 1998, the United States finally carried out its oft-repeated threat and launched a heavy military strike. After four hundred cruise missiles had hit Iraq, Saddam emerged from the smoke and ashes with his power apparently undiminished.

Unable to strike effectively, the United States has been quite willing to wound. The enforcement of economic sanctions has been the only instrument of policy toward Iraq pursued with consistency and vigor by successive U.S. administrations. While this weapon was deemed a "demonstrable success" in keeping Saddam weak, the real wounded were to be found among the Iraqi people. By 1998, four to five thousand children were dying every month because of sanctions. Alarmed that Saddam was turning the suffering of his people to his own propaganda advantage, the United States encouraged the introduction of the oil-for-food program. But by December 1998, after this program had been in operation for almost two years, half of all Iraqi children were still malnourished. More food was arriving, but widespread lack of clean water, a functioning sewage system, or electricity meant that Iraq had returned to a preindustrial age. Despite this exercise in what Pope John-Paul II has called "biological warfare" against the Iraqi people, the economic blockade has not weakened the regime's grip on power. Reliance on the official ration has, if anything, strengthened the government's control. Sanctions have proved ineffective in forcing Saddam Hussein to comply with United Nations resolutions and led to a tragedy of which the Western world is largely unaware. No one really knows how many have died as a result of sanctions so far, but reputable international bodies put the number of casualties among children alone at well over half a million. This is far higher than the death toll from the Gulf War and indeed approaches the benchmark for modern holocausts set in Rwanda and Cambodia.

Arguments against sanctions, of course, leave open the problem of the continued rule of the man against whom they are ostensibly aimed. The question "What do we do about Saddam?" is posed with increasing desperation during each new crisis between Iraq and the United States. Those who ask it usually want to be told that there is a simple formula for getting rid of the Iraqi leader. There are many who are willing to claim that, given enough willpower in Washington, his departure would not be so difficult to arrange by military coup or guerrilla warfare. But in practice the United States and its allies passed up their chance of getting rid of Saddam Hussein in 1991 when their armies stopped on the borders of Iraq. Nobody now shows any enthusiasm for a renewed U.S. military buildup followed by a ground assault—even if Saudi Arabia was prepared to lend its support for such a venture.

The more realistic question that should be asked is not how to get rid of Saddam but how to limit his ability to do harm. This has been the justification for Unscom and its weapons inspectors. They prevented Saddam Hussein from rebuilding his nuclear, chemical, and biological weapons programs (beyond the embryonic state in which they may presently exist) and using them against his neighbors. But the main victims of Saddam Hussein have always been ordinary people. It is they who have suffered through two wars.

They have seen their country destroyed. Ironically, it is they who are the primary target of sanctions.

Any visitor to Iraq knows that Iraqis blame sanctions and those who enforce them for much of their present misery. It is no less clear that bitterness and hatred of their ruler also runs deep. Saddam's downfall will come at the hands of his own people, independent of outside intervention—a fact of which he himself is well aware. He knows that the rage and hatred of the masses who for a few delirious days defaced his portraits and lynched his henchmen in March 1991 has not gone away. Sooner or later there will be a reckoning.

Notes

CHAPTER 1: SADDAM AT THE ABYSS

page 5   Statements: Foreign Broadcast Information Service, Near East Survey (FBIS, NES) 91 043, p. 33.

page 7   Informed opinion: see Colin Powell, *My American Journey,* (New York: Random House, 1995), p. 461.

page 7   "What is politics?": Interview with Dr. Hussain al-Shahristani, Tehran, 4/10/98.

page 7   "I consider myself to have died then": Interview with Abdel Karim Kabariti, Amman, 3/9/98.

page 7   Invasion decision from God: *Guardian,* London, 6/10/91.

page 8   Tariq Aziz: Interview by Andrew Cockburn with Zeid Rifai, Amman, February 1992.

page 8   Khairallah: Wafiq al-Samarrai, however, thinks that it was genuinely an accident, recalling that the weather was bad enough that day to tear the roof off his headquarters. Interview, London, 3/10/98.

page 8   "Only two divisional commanders": Pierre Salinger, *Secret Dossier: The Hidden Agenda Behind the Gulf War* (London: Penguin Books, 1991), p. 15.

page 8   "if I make a peace proposal": Ibid., p. 187.

page 9   "stay motionless": Faleh Jabber, in *Iraq Since the Gulf War, Prospects for Democracy,* Fran Hazelton, editor (London: ZED Books, 1994), p. 104.

page 9   "Las Vegas": Michael Gordon and Bernard E. Trainor, *The Generals' War* (New York: Back Bay Books, 1995), p. 215.

page 9   Racetrack: Patrick Cockburn, *Independent,* London, 1/17/91.

page 10   "Our main hobby": Interview in Baghdad, 1/16/91.

page 10   fear of nuclear strike: This point was made to Patrick Cockburn by several Iraqis in late January. By the time it was clear that this was unlikely to happen, there was no fuel for the refugees to return.

page 10   elephants: Personal observation, Patrick Cockburn, Baghdad, 2/17/91.

page 10   Saddam with the troops: Patrick Cockburn, Baghdad, 1/16/91.

page 10   "We knew all about these weapons": Interview with Brigadier Ali, London, 3/13/98.

page 11   "their ancient reputation for savagery . . .": Sir Arnold Wilson: *Loyalties: Mesopotamia 1914-1917, Vol. 1,* (Oxford: University Press, 1930), p. 136.

page 13   Captain Shirwan: Interview, Salahudin, Kurdistan, June 1991.

page 13   United States avoided researching casualties: Interview with former senior CIA official, Washington, 2/8/98.

page 13   "We didn't lose a single officer": Interview with Wafiq al-Samarrai, London, 6/2/98.

page 13   No casualties from Tulaiha: Interview with Hassan Hamzi, Tulaiha, July 1991.

page 15   "We were anxious to withdraw": Faleh Jabber in *Why the Intifada Failed,* Fran Hazelton, editor (London: Zed Books, 1993), p. 107.

page 16   The hotels: Patrick Cockburn, interviews in Basra, 4/22/91.

page 19   Outbreak of Kurdish intifada: Jonathan Randal, *After Such Knowledge, What Forgiveness?: My Encounters with Kurdistan* (New York: Farrar, Straus & Giroux, 1997), pp. 40–41.

page 19   "Haji Bush," Ibid., p. 45.

page 20   Hillah: Interview with Hussain al-Shahristani, Tehran, 4/10/98.

page 20   "At first we were a little crazy": Kanan Makiya, *Cruelty and Silence* (New York: W. W. Norton, 1993), p. 73.

page 21   Fatwa and bodies: ibid., pp. 74–75.

page 21   "Nobody knew what was going on": Interview with Sayid Majid al-Khoie, London, 6/2/98.

page 21   Jabber, op. cit., pp. 108–109.

page 22   "They swear by the Koran": Interview with Saad Jabr, London, 3/12/98.

page 22   Iranian behavior: Interview with Hussain al-Shahristani, Tehran, 3/10/98.

page 25   Intercepted message: Interview with Wafiq al-Samarrai, London, 3/10/98.

page 25 "one thousand dead in Basra": Interview and personal observation by Patrick Cockburn in Basra, 4/22/91.

page 26 Scenes in Kerbala and Najaf: Visit to Kerbala and Najaf, Patrick Cockburn, 4/15/91. Interview with General Abdul Khaliq Abdul Aziz, governor of Kerbala, and Abdul Rahman al-Dhouri, governor of Najaf. Both men emphasized Iranian involvement, showing some ammunition and a TNT charge that they said were of Iranian origin.

page 26 Al-Khoie meeting with Saddam, and Mohammed's interrogation: Interview with Sayid Majid al-Khoie, London, 6/2/98. Mohammed Taqi recounted these events to his brother, Sayid Majid, before he was killed in what his family is convinced was a government-arranged car accident between Najaf and Kerbala on July 21, 1994.

page 27 Al-Khoie presented by the government: Observed by Patrick Cockburn, Kufa, 4/15/91.

page 27 The place can be identified because at one point the film shows a road sign saying "Rumaytha."

page 28 "Only when you Kurds took Kirkuk": Interview with Hoshyar Zibari, London, 6/3/98.

page 28 Thousands of Kurdish volunteers: Interview with Massoud Barzani, Salahudin, May 1991.

page 29 Bullets: Interview with Wafiq al-Sammari by Andrew and Patrick Cockburn, London, 3/10/98.

page 29 "mistakes": Interview with Saad al-Bazzaz, former editor of *al-Jumaniya,* by Patrick Cockburn, London, 3/16/98.

CHAPTER 2: "WE HAVE SADDAM HUSSEIN STILL HERE"

page 31 "We didn't have a single": ABC News, *Peter Jennings Reporting,* "Unfinished Business: The CIA and Saddam Hussein," 6/26/97.

page 32 Escape of Republican Guard: Colonel James G. Burton, USAF (retired), June 1993, "Pushing Them out the Back Door," *United States Naval Institute Proceedings.*

page 33 advance on Baghdad: Michael R. Gordon and Bernard E. Trainor, *The Generals' War* (New York: Back Bay Books, 1995), p. 452.

page 33 "The White House was terrified": Interview with Ambassador Charles Freeman, Washington, 3/31/98.

page 33 "no political overtones": Telephone discussion with James Akins, 5/28/98.

page 34   Saddam at top of target list: Rick Atkinson, *Crusade: The Untold Story of the Persian Gulf War* (New York: Houghton Mifflin, 1993), p. 272.

page 34   "focus of our efforts": *Washington Post,* 9/16/90.

page 34   "We don't do assassinations": ABC News, *Peter Jennings Reporting,* "Unfinished Business: The CIA and Saddam Hussein," 6/26/97.

page 36   Agent report on shelter: Atkinson, op. cit., p. 276.

page 36   Schultz forbids contacts: Senate Foreign Relations Committee report, "Civil War in Iraq," (Washington, D.C.: Government Printing Office), 5/1/91.

page 36   "stupid stuff": Interview with Peter Galbraith; Washington, D.C.; 5/30/98.

page 37   "get rid of Saddam": Senate Foreign Relations Committee, op. cit.

page 38   "I did have a strong feeling": Gordon and Trainor, op. cit., p. 517. (Authors' emphasis in quotation.)

page 39   "the Iranian occupation": Ibid., p. 516.

page 39   CIA gives arms: Former American diplomat who asked not to be named; interviewed in Washington, D.C., 5/29/98.

page 40   "never our goal": Gordon and Trainor, op. cit., p. 517.

page 40   "flimsy construction": Interview with Ambassador Freeman, 3/31/98.

page 40   President compared to Lincoln: Mary McGrory, *Washington Post,* 3/26/91.

page 40   White House meeting: *Washington Post,* 3/27/91.

page 41   "Saddam will crush": *Washington Post,* 3/29/91.

page 41   "Not that I know of": Question and answer session with reporters at Bethesda Ward Hospital, Maryland, 3/27/91. Transcription from the public papers of George Bush, www.csdl. TAMU.edu/BushLib/

page 41   "Why are you so worried about the Shia": Interview with Sayid Majid al-Khoie, London, 6/2/98.

page 42   "Do I think": presidential press conference, 4/16/91.

page 43   Pickering, Gates: *Los Angeles Times,* "U.S. Sanctions Threat Takes UN by Surprise," 5/9/91, p. A10.

page 50   "smart but not wise": Interview with Abdul Karim al-Kabariti, 3/9/98.

page 52   Fadlallah a follower of al-Khoie: Olivier Roy, *The Failure of Political Islam* (London: Penguin, 1995), p. 57.

CHAPTER 3: THE ORIGINS OF SADDAM HUSSEIN

page 58 A week to travel to Basra: Hanna Batatu, *The Old Social Classes and the Revolutionary Movements of Iraq* (Princeton, NJ: Princeton University Press, 1978), p. 16

page 59 thirteen days to Basra: Norman F. Dixon, *On the Psychology of Military Incompetence* (London: Jonathan Cape, 1976), p. 103.

page 59 Rebellion in Najaf: Batatu, op. cit., p. 19.

page 61 "To depend on the tribe": Ibid., p. 21.

page 62 British defeat at Kut: David Fromkin, *A Peace to End All Peace* (London: André Deutsch, 1989), pp. 200–203.

page 62 British cemetery in Kut now a swamp: Personal observation by Patrick Cockburn, April 1998.

page 63 "the antithesis of democratic government": H.V.F. Winstone, *Gertrude Bell* (London: Jonathan Cape, 1978), pp. 215–216.

page 63 three quarters of the population were tribal: Fromkin, op. cit., pp. 449–450.

page 63 "a flabby serpent": Batatu, op. cit., p. 14.

page 63 "The bottom seems to have dropped out": General Aylmer Haldane, *The Insurrection in Mesopotamia* (Cambridge: Allborough Publishing, 1992), p. 37.

page 63 "Since you took Baghdad": Winstone, op. cit., p. 222.

page 64 63,000 rifles: Elie Kedourie, *Politics in the Middle East* (Oxford: Oxford University Press, 1992), p. 195.

page 64 "The bottom seems to have dropped out": *Selected Letters of Gertrude Bell,* edited by Lady Bell, (London: Ernest Bell, 1927), Vol. II, p. 489. She was writing on June 14, less than three weeks before the uprising.

page 64 "The Arab is most treacherous": Haldane, op. cit., p. 36.

page 64 Shia-Sunni unity: Batatu, op. cit., p. 23.

page 65 "They now know what real bombing means": David McDowall, *A Modern History of the Kurds* (London: B. Tauris, 1997), p. 180.

page 65 "I am very doubtful": Haldane, op. cit., p. xiii.

page 65 T. E. Lawrence and poison gas: Ibid., p. vi.

page 66 "There is still": Batatu, op. cit., pp.25–26.

page 67 Death of royal family: Ibid., op. cit., p. 801.

page 67 "Iraq today": Said K. Aburish, *A Brutal Friendship: The West and the Arab Elite* (London: Victor Gollancz, 1997), p. 135.

page 67 Iraqi poets winning prizes: Interview with Faleh Jabber, London, 6/24/98. Poets who wrote that the comparison was to Saddam's advantage qualified for larger prizes.

page 69   "To talk like a Tikriti": Gavin Young, *Iraq: Land of Two Rivers* (London: Collins, 1980), p. 98

page 69   "One of my uncles": Fuad Mattar, *Saddam Hussein: The Man the Cause and the Future* (London: Third World Center, 1981), p. 228

page 70   safer in prison than on the streets: Ibid., pp. 31–32

page 70   Escape from jail: Ibid., p. 46

page 70   "They gave him a pistol": Ibid., p. 31.

page 71   "My headmaster told me": Interview with Dr. Abdul Wahad al-Hakim, "The Mind of Saddam Hussein," WGBH, *Frontline*, Boston, 2/26/91.

page 71   Saddam pays blood money: Interview with Faleh Jabar, London, 6/25/98.

page 71   "A Baathi would have looked in vain": Batatu, op. cit., p. 1014.

page 72   "a very superficial wound": Interview with Dr. Tahsin Muallah, WGBH, *Frontline*, Boston, 2/26/91.

page 73   "It was like you see in the movies": *Independent*, 3/31/98.

page 73   Saddam's escape: Mattar, op. cit., pp. 33–43.

page 73   "We helped him go": Interview with Abdel Majid Farid, London, 6/2/98.

page 74   Saddam and the bar in Cairo: *New York Times*, 10/24/90.

page 74   "A great victory": Interview with James Critchfield; Washington, D.C.; 4/10/91.

page 74   "CIA train": Aburish, op. cit.

page 75   Qassim's body: Kedourie, op. cit., p. 320.

page 75   "military aristocracy": Interview with Faleh Jabber, 6/24/98.

page 76   "exactly as we used to run Tikrit": Interview with Kamran Karadaghi, Iraqi journalist, London, 1997.

page 76   Saddam's health and taste for Portuguese rosé: Interview with Wafiq al-Samarrai, London, 3/10/98.

page 76   "He will go back": Interview with Kamran Karadaghi, 6/6/98.

page 78   Trial of "conspirators": Transcript of videotape shown on WGBH, *Frontline*, Boston, 2/26/91.

page 78   Body returned in a truck: Iraqi source, name withheld by request.

page 80   "that mummy": Mattar, op. cit., pp. 130–135.

page 80   DIA report: *Independent*, London, 12/12/92

page 81   CIA aid to Saddam: Interview with Wafiq al-Samarrai, London, 3/10/98.

page 82   "War doesn't mean just tanks": Pierre Salinger, *Secret Dossier: The Hidden Agenda Behind the Gulf War* (London: Penguin Books, 1991) p. 31.

page 82   Iraq's debt in 1990: Barry Rubin and Amatzia Baram, editors, *Iraq's Road to War* (New York: St. Martin's Press, 1993), pp. 70–83.

page 83   "The ground bursts open": Samir al-Khalil, *The Monument: Art, Vulgarity, and Responsibility in Iraq* (London: André Deutsch, 1991) p. 2.

page 83   "We agreed with the American side": Salinger, op. cit., pp. 239–41.

page 83   "You are a force for moderation": Ibid., p. 65.

page 84   "When the king got back to Amman": Interview with Abdul Karim al-Kabariti, Amman, 3/9/98.

page 84   "burn half your house": Rubin and Baram, op. cit., p. 12.

page 84   "The Saudis want to weaken us": Salinger, op. cit., p. 65.

page 85   Saddam's original plan: Milton Viorst, "Interview with Tariq Aziz," *The New Yorker,* 6/24/91.

CHAPTER 4: SADDAM FIGHTS FOR HIS LONG ARM

page 89   Experiences of Dr. Hussain al-Shahristani: Interviews by Andrew Cockburn with Dr. al-Shahristani, Tehran, 4/10/98, 4/16/98.

page 90   Information on the nuclear program: "The Implementation of United Nations Security Council Resolutions Relating to Iraq." Report by the Director General, International Atomic Energy Agency General Conference, August 12, 1996.

page 92   Gas attack on Tehran: Interview with Wafiq al-Samarrai, London, 3/13/98.

page 93   "the United States will not tolerate": Statement by Press Secretary Fitzwater on President Bush's letter to President Saddam Hussein of Iraq, released by the White House, 1/12/91.

page 93   Boomer's revelation: Interview with General Walter Boomer by Leslie Cockburn, Saudi Arabia, 9/90.

page 94   "which shall carry out": Security Council Resolution 687, paragraph 9(b)(1).

page 96   "Temporary measure": Interview with Wafiq al-Samarrai, London, 3/12/98.

page 98   "I thought it should be over quickly": Interview with Ambassador Rolf Ekeus; Washington, D.C.; 2/9/98.

page 98   "We were very, very skeptical . . . Ekeus's open-minded approach": Interview with former senior CIA official; Washington, D.C.; 2/6/98.

page 102   Meeting with Tariq Aziz and Hussein Kamel: Interview with Rolf Ekeus, 2/9/98.

page 102   High-level committee: Presentation by Ambassador Richard Butler to the UN Security Council, 6/3/98.

page 103   Digging up garden: Ibid.

page 105   CIA official on Saddam's plan for weapons program: Interview with former senior CIA official; Washington, D.C.; 2/16/98.

page 106   Group with placards: *Independent*, London, 9/26/98.

page 110   ". . . the Commission is approaching": Report to the Security Council, S/1994/138, 10/7/94.

page 111   Rockville, Maryland: *Washington Post*, 11/21/97.

page 112   Spertzel, growth media, Aziz explanation: Interview with Rolf Ekeus; Washington, D.C.; 6/16/98.

CHAPTER 5: "IRAQIS WILL PAY THE PRICE"

page 114   No food in the garbage: Observation by Patrick Cockburn, 7/25/91.

page 115   World Bank equates Iraq with Greece: Anthony Cordesman and Ahmed S. Hashim, *Iraq: Sanctions and Beyond* (Boulder, CO: Westview, 1997), p. 127.

page 115   Iraqi chicken: Interview with Doug Broderick, Catholic Relief Services, Baghdad, 9/7/91.

page 115   Oil revenues: Peter Boone, Haris Gazdar, and Athar Hussein, "Sanctions Against Iraq: The Costs of Failure" (New York: Center for Economic and Social Rights, November 1997), p. 8.

page 115   2000 percent food price increase: *Middle East Contemporary Survey*, volume XV (Boulder, CO: Westview, 1991), p. 437.

page 115   Selling rags as clothes: Personal observation by authors, July 1991.

page 116   Scene at church and aid worker comment: Personal observation by authors and interview with Doug Broderick of Catholic Relief Services, Baghdad, July 1991.

page 120   Nuha al-Radi: Nuha al-Radi, *Baghdad Diaries* (London: Saqi Books, 1998), pp. 59–60.

page 122   Blunt scissors: "Unsanctioned Suffering: Human Rights Assessment of United Nations Sanctions on Iraq" (New York: Center for Economic and Social Rights, May 1996).

page 122   600 percent inflation: *Independent*, London, 4/22/91.

page 123   Ahtisaari: "Report of the United Nations Mission to Assess Humanitarian Needs in Iraq," March 10–16, 1991, led by Martti Ahtisaari, Undersecretary General for Administration and Management, excerpts in Middle East Report, May–June 1991, p. 12.

page 123   Aga Khan: *Independent*, London, 7/20/91.

page 123 Rations provide 53 percent of basic needs: "Unsanctioned Suffering," op. cit., p. 986.

page 123 "The system is highly equitable": Ibid., p. 18.

page 124 Kroll: *Los Angeles Times,* 3/23/91, p. A–12.

page 125 Episode in Amman office: Leslie and Andrew Cockburn, "Saddam's Best Ally," *Vanity Fair,* August 1992.

page 125 Forty-two merchants executed: Cordesman and Hashim, op. cit., p. 141.

page 125 "because of the lack of machinery": Interview by Patrick Cockburn with Khalid Abdul Munam Rashid, 10/17/95.

page 125 40 percent in agriculture: Boone, Gazdar, and Hussein, op. cit., p. 25.

page 126 Price of grain: Ibid., pp. 17–18.

page 128 Amputations: Middle East Contemporary Survey, op. cit., pp. 337–339.

page 129 Blank check: Leslie and Andrew Cockburn, op. cit.

page 132 Sewage system: Information supplied by Abdullah Mutawi of the Center for Economic and Social Rights, New York, who was on both the Harvard and CESR trips in 1991 and 1996.

page 132 Deaths from drinking contaminated water: Interview with Dr. Nada al-Ward, WHO, Baghdad, 6/20/98.

page 132 "quarter of the children are suffering from malnutrition": *Independent,* London, 10/14/85.

page 132 Baghdad study of children: *The Lancet,* 346, 12/2/95. The study was done by Sarah Zandi and Mary Sith Fawzi on August 23–28, 1995.

page 135 "semi-starvation diet": "The Health Conditions of the Population in Iraq Since the Gulf Crisis" (Geneva: WHO, March 1996,) p. 8.

page 136 Dr. Obousy: Interview by Patrick Cockburn with Dr. Deraid Obousy, Baghdad, 4/19/98.

page 137 576,000 dead children: *New York Times,* 12/1/95.

page 137 90,000 dying every year: "Unsanctioned Suffering," op. cit., p. 20.

page 138 Madeleine Albright: CBS News, *60 Minutes,* 5/12/96.

page 138 Greece and Mali: Greece comparison from Cordesman and Hashim, op. cit., p.127 Mali comparison from *The Lancet,* 346, 12/2/95.

page 139 Prince Khalid: *New York Times,* 12/14/95.

CHAPTER 6: UDAY AND THE ROYAL FAMILY

page 142 Scene in National Restaurant: Leslie and Andrew Cockburn, "Saddam's Best Ally" *Vanity Fair,* August 1992.

page 143 Saddam pictured darning: Fuad Matar, *Saddam Hussein: The Man, the Cause, and the Future* (London: Third World Center, 1981), p. 251.

page 143 "The higher nobility": For an excellent chart showing the relationships among Saddam Hussein, the Ibrahims, and the Majids, see Faleh A. Jabber, "Batailles des clans de l'Irak," *Le Monde Diplomatique* (September 1966).

page 144 Ali Hassan al-Majid's health: Interview by Andrew and Patrick Cockburn with General Wafiq al-Samarrai, London, 4/12/98.

page 145 "What is this exaggerated figure": Jonathan Randal, *After Such Knowledge, What Forgiveness?* (New York: Farrar, Straus & Giroux, 1997), pp. 212–214.

page 145 Hussein at Kerbala: Hussein Kamel's real crime at Kerbala, apart from the mass execution of rebels, was to destroy much of the old city. There are two shrines in Kerbala, one containing the tomb of Imam Hussein and the other his brother al-Abbas, the founding martyrs of the Shia faith, who died in battle in A.D. 680. The shrines are about five hundred yards apart. In 1991, the Iraqi army, led by Hussein Kamel, systematically destroyed all the buildings between the shrines. The site is now a public garden.

page 146 Teetotaler Hussein Kamel: Interview with Abdul Karim al-Khabariti, former Jordanian foreign and prime minister, Amman, 9/3/98.

page 146 "did not go to the office": Hussein Kamel, press conference, 8/12/98, reported on *BBC Summary of World Broadcasts,* 8/14/98.

page 147 Barzan: Interview with Barzan Ibrahim al-Tikriti, *al-Hayat,* translated in *Mideast Mirror,* 8/31/95.

page 147 "only legitimacy": Ibid.

page 149 "treacherous and perfidious people": Saddam Hussein, message to the Iraqi people, FBIS, 8/30/92, p. 16.

page 150 Saddam apologizes for land reforms: *Le Monde Diplomatique,* September 1996.

page 150 Faleh Jabber: "Why the Intifada Failed: Faleh Jabber," *Iraq Since the Gulf War,* Fran Hazelton, editor (London: ZED Books, 1994), p. 115.

page 151 He spoke about this when he was sixteen: Matar, op. cit., p. 16.

page 151 "I did my SATs": Leslie and Andrew Cockburn, op. cit.

page 151 Outings to the torture chamber: Information supplied by Charlie Glass, ABC News, from interview with General Hassan al-Naquib, 3/21/91.

page 152 Uday goes into action: Interview, General Wafiq al-Samarrai, London, 3/12/98.

page 154 Latif Yahia the double's account of the Jajo killing: Latif Yahia and Karl Wendl, *I Was Saddam's Son* (New York: Arcade, 1997), pp. 162–173.

page 155 King Hussein as counselor: Interviews with former close adviser to king, Amman, 2/21/93.

page 156 Abdul: Interview with "Abdul," Washington, D.C., 8/20/98.

page 161 Terrorist bombs in Baghdad: *Middle East Contemporary Survey,* vol. XVIII, 1994, p. 327, citing *Babel,* 2/2/94.

CHAPTER 7: INTRIGUE IN THE MOUNTAINS

page 164 Kuwait crisis, Clinton address: "Clinton ups Heat on Iraq," *Chicago Tribune,* 10/11/94.

page 165 Offhand remark by Clinton: Thomas Friedman, *New York Times,* 1/15/95.

page 166 Key Bridge Marriott: ABC News, *Peter Jennings Reporting,* "Unfinished Business: The CIA and Saddam Hussein," 6/26/97.

page 167 "the capability": Ibid.

page 169 "The way the names were depicted": Interview with former CIA official; Washington, D.C.; 3/19/98.

page 170 Fallout from the Ames case: Tim Weiner, David Johnston, and Neil Lewis, *Betrayal: The Story of Aldrich Ames, an American Spy* (New York: Random House, 1995), pp. 285–287, and David Wise, *Nightmover* (New York: HarperCollins, 1995), pp. 310–311.

page 171 "What I wanted them to do": ABC News, op. cit.

page 174 CIA recruits Nuri; "work separately": Ibid.

page 175 PKK: The acronym comes from the Kurdish name "Partei Karkaren Kurd," which translates as Kurdistan Workers' Party.

page 176 CIA support for Barzani: *CIA: The Pike Report* (London: Spokesman Books, 1977) p. 197.

page 177 Saddam speech, 3/16/91: FBIS-NES–91–052, p. 28.

page 177 Kurdish politics: *A Modern History of the Kurds* by David MacDowall (London: I. B. Tauris, 1997) is an indispensable guide to the fractured history of the Kurds. For the shifting alliances discussed above, see pp. 343–354.

page 177 "They are obsessed": MacDowall, op. cit., p. 385.

page 178 INC militia numbers: *Middle East Contemporary Survey*, volume XVIII, 1994, p. 348.

page 179 May 1994 fighting: MacDowall, op. cit., p. 386.

page 182 Chalabi borrows money: Interview with former INC official, London, 3/12/98; interview with former CIA official, Washington, 6/20/98.

page 183 "The INA was as leaky as a sieve": Interview with former CIA official; Washington, D.C.; 8/20/98.

page 183 "I liked Bob": Interview with Hoshyar Zibari; Washington, D.C.; 3/19/98.

page 186 "I told him": *Los Angeles Times*, 2/15/98.

page 186 "Wafiq, who we had been paying": Interview with Hoshyar Zibari; Washington, D.C.; 3/19/98.

page 188 Nuri flies to Washington: ABC News, op. cit.

CHAPTER 8: DEATHS IN THE FAMILY

page 193 "Ten days before we decided to travel": Hussein Kamel, press conference, 8/12/95, *BBC Survey of World Broadcasts*, 8/14/95.

page 193 Kamel collects money: He later confided this to Rolf Ekeus. Interview with Rolf Ekeus; Washington, D.C.; 6/16/98.

page 194 "We knew he had crossed the border": Interview with Abdul Karim al-Kabariti, Amman, 3/9/98.

page 195 "Don't let His Majesty shake hands": Ibid.

page 196 "My daughters spend time with them": Ibid.

page 196 King's interview: Associated Press, 8/14/95.

page 196 "sinking ship": Jim Hoagland, *Washington Post*, 8/17/95.

page 197 "Between him and the Kurds": Barzani interview, *al-Hayat*, translated in *Mideast Mirror*, 8/31/95.

page 198 Tanous: Interview with a source close to the Jordanian government, Amman, 10/16/95.

page 198 "boiling with hatred": Interview with Rolf Ekeus, New York, 4/24/97. Quoted in *One Point Safe* by Andrew and Leslie Cockburn (New York: Anchor Doubleday, 1997), p. 215.

page 199 "We are incompetent": Interview with Rolf Ekeus; Washington, D.C.; 6/16/98.

page 200 Ekeus's negotiations with the Iraqis immediately before and after the defection of Kamel: Report of the Executive Chairman of the Special Commission to the UN Security Council, S/1995/864, 10/11/95.

page 200 Purging of files: Presentation by Ambassador Richard Butler to UN Security Council, 6/3/98.

page 201 Uday's Fedayeen: *Independent*, London; 8/31/95. They were seen by diplomats traveling on the road between Amman and Baghdad.

page 201 "there is not a single street": Hussein Kamel, press conference, *BBC Survey of World Broadcasts*, 8/14/95.

page 201 "People held parties": *Independent*, London; 8/16/95.

page 201 Telephone link to Jordan: *BBC Survey of World Broadcasts*, 8/14/95. The correspondent of the Egyptian news agency MENA discovered that he could call direct between the two capitals.

page 201 Saddam denounces Kamel: *BBC Survey of World Broadcasts*, 8/14/95.

page 202 "His family has unanimously decided": Iraqi TV, 8/12/95. The station broke into normal programming to make the announcement.

page 202 Kamel's illiterate letter: *International Herald Tribune*, 9/18/95.

page 203 Saddam's raid on Uday's garage: *Independent*, London; 10/12/95. Despite the raid on the Olympic committee building, lights continued to shine in its offices at night. The burning car story could not be checked, but nobody admitted to seeing the smoke from the burning buildings.

page 203 Health of Saddam Hussein: Interview with Wafiq al-Samarrai, London, 3/12/98.

page 203 "O lofty mountain!" *BBC Survey of World Broadcasts*, 8/14/95.

page 203 "Victory is sweet": *Independent*, 10/14/95.

page 203 British poll: Philip Willard Ireland, *Iraq: A Study in Political Development* (London: Jonathan Cape, 1937), p. 332.

page 204 Medical checkup: *Washington Post*, 2/24/96.

page 205 Ekeus still visiting Kamel: Interview with Rolf Ekeus; Washington, D.C.; 6/16/98.

page 205 "The girls will have to stay here": Interview with Abdul Karim al-Kabariti, Amman, 3/9/95.

page 206 "Higher Council": Anthony H. Cordesman and Ahmad S. Hashim, *Iraq: Sanctions and Beyond* (Boulder, CO: Westview Press, 1997), pp. 68–69.

page 207 Suck him dry: *BBC Summary of World Broadcasts*, 8/14/95.

page 208 "the leadership took a decision": *Washington Post*, 2/24/96.

page 209 Saddam demands brothers divorce wives: Interview with Abbas Jenabi, Associated Press, 10/1/98.

page 209 Divorce: *Independent*, London; 9/24/96, citing Iraqi News Agency.

page 210 "We have cut off the treacherous branch": *Le Monde Diplomatique,* 9/96.

page 210 "Had they asked me": Cordesman and Hashim, op. cit., p. 27.

CHAPTER 9: "BRING ME THE HEAD OF SADDAM HUSSEIN"

page 213 Amneh's tape: The existence of Abu Amneh's tape was revealed by Patrick Cockburn in *The Independent,* London, 3/26/98.

page 214 Arrest of bombers of INC building: Interview with Ghanim Jawad, senior INC official, London, 9/4/98. Involvement of Amneh in bombing: Interview with a senior INC official, 3/14/98.

page 214 "Bureaucrats in the CIA": Interview with former CIA official; Washington, D.C.; 6/18/98.

page 216 Banner with Churchill quotation: Described in the *Washington Post,* 9/15/96, though without naming the official. Mattingly was a forthright individual, famous in the agency for an incident earlier in his career when he had been serving as acting CIA station chief at the U.S. embassy in Turkey. The ambassador was a somewhat eccentric individual named Strausz Huppé. One morning the ambassador read in the paper that Kurt Waldheim, the former Nazi serving as UN Secretary General, was coming to Ankara. At the morning staff meeting, he launched into a tirade on the subject of Waldheim's iniquities, climaxing with a question to Mattingly: "Mattingly, can you kill him?" To which Mattingly immediately shot back: "Yes, sir, I can. But I won't."

page 217 "There should be a rule": Interview with a former senior CIA official; Washington, D.C.; 2/6/98.

page 217 "He would be in a meeting": Interview with a former very senior CIA official; Washington, D.C.; 2/28/98.

page 217 "Deutch mistrusted people": Interview with a former senior CIA official; Washington, D.C.; 3/5/98.

page 218 Tighter and more focused: *Washington Post,* 9/15/96.

page 218 "Deutch recruited subordinates": Interview with a former senior CIA official; Washington, D.C.; 2/6/98.

page 219 "we were significantly affected": Interview with a former senior CIA official; Washington, D.C.; 2/10/98.

page 219 Turki's secret trip: Anthony H. Cordesman and Amad S. Hashim, *Iraq: Sanctions and Beyond* (Boulder, CO: Westview, 1997), p. 194.

page 219 "I wasn't given any briefings": Telephone interview with a former CIA official, 9/17/98.

page 219 "It is my understanding": Interview with a former CIA official, Arlington, 3/19/98.

page 221 "Not even the chief of staff": Telephone interview with Harold Ickes, 9/18/98.

page 221 "He was much more gung ho": Interview; Washington, D.C.; 9/21/98.

page 221 "Deutch came back from a meeting": Interview; Washington, D.C.; 4/6/97.

page 222 Naji murder: Andrew and Leslie Cockburn, *One Point Safe* (New York: Anchor Doubleday, 1997), p. 200.

page 223 Al-Khazraji's statement: *al-Hayat*, 4/4/96, as translated in *BBC Summary of World Broadcasts*, 4/4/96.

page 223 Al-Khazraji dumped by CIA: Interview with an Iraqi opposition source; Washington, D.C.; 2/19/98.

page 224 Telephone eavesdropping system: Sean Boyne, "Inside Iraq's Security Network," *Jane's Intelligence Review*, vol. 9, no. 7, 7/1/97.

page 225 Accord press conference: Press statement on behalf of Dr. Alawi, Secretary General of the Iraqi National Accord, issued by the INA, 2/18/96 (one of the few press announcements remaining on the INA Website as of September 1998).

page 225 Alawi on CNN: 3/2/96. Posted on CNN Website at 11:55 P.M. EST, 3/2/96.

page 225 "unqualified success": *Washington Post*, 9/29/96.

page 225 Cockburn article in the *Independent*: Patrick Cockburn, "Clinton Backed Baghdad Bombers," *Independent*, London, 3/26/96.

page 225 Antiterror conference: Ibid.

page 226 Nuri leaves Iraq: *Al-Quds,* 7/18/96, 7/22/96.

page 226 "all penetrated by Iraqi security": Interview with Abdul Karim al-Kabariti, Amman, 3/9/98.

page 226 News of Iraqi penetration and Chalabi's trip to Washington: This account is drawn largely from INC and other Iraqi opposition sources and is confirmed by CIA sources.

page 227 Alawi interview: *Washington Post*, 6/23/96. Picked up by wires, see, for example, Arab Press Service Organization, 6/23/96.

page 228 June 20: A press release from the Iraqi National Accord, "Attempted Coup in Iraq," 7/11/96, dates the first arrests to June 20. Other sources suggest they began six days later.

page 228 Names of those arrested: Drawn from a letter from Ghanim Jawad, an exceptionally well-informed member of the non-Accord opposition, to Amnesty International, 11/3/96.

page 229   Accord press release: "Attempted Coup in Iraq: Update, Death During Interrogation," 7/12/96.

page 230   Deutch's statement: ABC News, *Peter Jennings Reporting*, "Unfinished Business: The CIA and Saddam Hussein," 6/26/97.

CHAPTER 10: SADDAM MOVES NORTH

page 231   "take a whack at his prestige": Interview with a former senior CIA official; Washington, D.C.; 2/6/98.

page 233   Hamilton Road and importance of Gali Ali Beg: *Independent*, London, 7/6/96.

page 233   Vickers machine gun: Interview with former Ambassador Bill Eagleton, one of the most knowledgeable Americans on the subject of the Kurds, who encountered the Sourchi while stationed in the U.S. embassy in Baghdad in the 1950s.

page 233   Luxurious villas: Personal observation, August 1991.

page 234   "they either tell Zayed to go away, or": Interview with Hoshyar Zibari; Washington, D.C.; 9/7/98.

page 234   "My father was expecting Massoud Barzani to come to lunch": Interview with Jahwar al-Sourchi, London, 9/8/98.

page 234   Sourchi homes leveled, ducks wandering through wreckage: Patrick Cockburn's visit to Kalaqin, 9/15/96.

page 235   "Many people talk about Massoud": Interview with Kamran Karadaghi, London, 9/7/98.

page 236   Talabani betrays Iranian Kurds: David MacDowall, *A Modern History of the Kurds* (London: I. B. Tauris, 1997), p. 451. To do him credit, Talabani also tipped off the KDPI that the Iranians were coming.

page 236   Warning to the NSC: *Independent*, 9/6/96.

page 237   Peshmerga numbers: An estimate by exceptionally well-informed Iraqi opposition observer Ghanim Jawad in an interview in London, 9/8/98.

page 237   Attack "backed by howitzers": *Independent*, London, 8/22/96.

page 238   "We request the U.S." *Independent*, London, 9/6/96.

page 239   Talabani "promised full cooperation": Robert Pelletreau, *al-Hayat*, 8/2/98.

page 240   Ahmad Allawi began to hear reports: Interview, 9/7/98.

page 243   "It is very possible that a lot of INC people were killed": ABC News, *Peter Jennings Reporting*, "Unfinished Business: The CIA and Saddam Hussein," 6/26/97.

page 243   Clinton statement: *Chicago Tribune*, 9/1/96.

page 243   Perry, "My judgment is": *International Herald Tribune,* 9/9/96.

page 243   Fear of being seen as Iran's ally: *Washington Post,* 9/8/96.

page 244   "We have choked Saddam Hussein in the south": *International Herald Tribune,* 9/9/96.

page 244   Deutch's testimony: *Washington Post,* 9/20/96.

page 244   KDP secures release of Islamic prisoners: Interview with members of the Islamic Movement of Kurdistan, Arbil, 9/14/96.

page 245   Public record of Dole's encounter with Saddam: The Iraqis had malignly released a transcript after the invasion of Kuwait.

page 245   Al-Kabariti calls the Americans: Interview with Abdul Karim al-Kabariti, Amman, 3/9/96.

page 246   Press criticism: *Washington Post,* 9/9/96.

page 246   Official to *Washington Post:* 9/10/96.

page 246   "veracity": Telephone interview with Ahmad Chalabi, 9/23/98.

page 247   Scene at MCC building: Interviews by Patrick Cockburn, Zakho, 9/14/96.

page 247   Mines advisory group: Interview with members of the advisory group by Patrick Cockburn, Diyana, 9/16/96.

CHAPTER 11: UDAY TAKES A HIT

page 253   Account of the ambush: This account is based on a detailed interview by Patrick Cockburn with Ismail Othman, a member of the group that organized the ambush, in London in the summer of 1997.

page 253   Jordanian diplomat: Radio Monte Carlo, Randah Habib in Amman, 12/13/96.

page 253   "slightly wounded": FBIS, NES 96 242. Agence France Presse (AFP), 12/16/96, quoting Iraqi News Agency.

page 253   Stock market crashes: FBIS, 12/15/96.

page 254   Journalists slaughter sheep: Baghdad Radio, 12/15/96.

page 254   Saddam orders care for others: AFP, 12/16/96.

page 254   Two thousand arrests: Voice of Rebellious Iraq (Shiite opposition), FBIS, NES 96 242, 12/15/96.

page 254   Sabawi and Watban: *Al-Sharq al-Awsat,* 12/14/96.

page 254   Uday suspects his father: Interview with a well-connected Iraqi source; Washington, D.C.; 11/20/98.

page 258   "It was you and Hussein Kamel": *Al-Wasat,* London, 3/12/97.

page 260   Battle at al-Kreeat: There is a dramatic and somewhat fanciful story about the fight at al-Kreeat in the Jordanian magazine *Sawt al-*

Marah on February 19, 1997. It says that it began after five gunmen had made a renewed attempt to kill Uday in the Ibn Sina hospital. It failed, and four of them were killed. The fifth was traced to al-Kreeat, which was attacked by a special force led by Uday! It says that seventy defenders were killed or captured, as well as four soldiers.

page 260   Saddam's bedside speech: *Al-Wasat,* London, 3/12/97.

page 261   Barzan resigns: *Al-Quds al-Arabi,* London, 9/23/98.

CHAPTER 12: ENDGAME

page 263   "price is worth it": CBS News, *60 Minutes,* 5/12/96.

page 265   "Not much is unknown": Report by the Secretary General, 4/11/97, S/1997/301.

page 265   King Fahd: Interview with a Western diplomatic source; Washington, D.C.; 10/10/97.

page 266   First concealment inspection: Unscom Report, 10/11/96, S/1996/848.

page 267   Concrete pillars: Ibid.

page 267   Russian spy: Interview with former Unscom official; Washington, D.C.; November 1997.

page 268   "miserable spy": Saddam Hussein, speech, 7/17/97, reported FBIS, 7/22/97.

page 268   "It is the Special Republican Guards": Interview with Rolf Ekeus; Washington, D.C.; 6/16/98.

page 270   Shell game: *Haaretz,* 9/29/98.

page 271   Helicopter, highest authority: Unscom Report, 10/6/97, S/1997/774.

page 271   Saddam's statement from RCC: Iraq TV News, 6/22/97.

page 272   Ritter gives U-2 photos to Israel: *Washington Post,* 9/29/98.

page 272   "gravest crisis of [Bill Clinton's] presidency": *Time* magazine, 11/24/97.

page 272   "head shot": *New York Times* columnist Thomas Friedman, quoted in *Time* magazine, op. cit.

page 272   "stunned": *Washington Post,* 3/1/98.

page 274   Ritter pulled back: *Washington Post,* 8/27/98.

page 275   Castor beans: Jim Hoagland, *Washington Post,* 2/11/98.

page 275   Bahrain: FBIS, 2/21/98.

page 276   Town hall meeting: 2/18/98. Text released by Department of State, 2/20/98.

page 277   "You can't go": *Washington Post,* 3/1/98.

page 277 Annan statement: *Washington Post*, 2/24/98.

page 278 Saddam on tour: Iraqi TV, 3/17/98, as reprinted in the online newsletter *Iraq News*.

page 278 "appeasement": Senator Trent Lott, *Washington Post*, 2/26/98. Senator John D. Ashcroft (Republican from Missouri) summed up the prevailing mood of his party when he declared that "U.S. foreign policy ought not to be subjected to Kofi Annan or written at the United Nations. And as long as I have a voice, America will not sacrifice another ounce of her sovereignty to the architects and acolytes of a one-world government" (*Washington Post*, 3/4/98).

page 278 Chalabi testimony: Senate Foreign Relations Committee, Sub-committee Hearings on the Middle East, 3/2/98.

page 279 Embezzlement question a plant: Jim Hoagland, "From Pariah to Iraq's Hope," *Washington Post*, 3/5/98.

page 279 Congress votes money: H.R. 3579 Sec. 2005, 4/30/98, *Iraq News*, 5/1/98.

page 279 Kurdish leaders considered INC defunct: Interview with Jalal Talabani, London, 6/6/98. Interview with Hoshyar Zibari, Washington, 3/16/98.

page 279 Sabotage plan: *New York Times*, 2/26/98.

page 282 Three nuclear weapons: Senate Foreign Relations and Armed Services Committees Hearings, 9/3/98.

page 282 *Haaretz*, 9/29/98.

page 283 Scott Ritter: Interview with Ritter in the *New York Post*, 12/17/98.

page 284 Message from al-Shahristani: E-mail from Dr. al-Shahristani to Andrew Cockburn, 12/18/98.

page 285 UNICEF: Nutritional status survey at primary health centers during polio national immunization days (PNID) in Iraq, March 14–16, 1998. Made available to authors by UNICEF office in Baghdad. The actual figures were: April 1997: Underweight—24.7 percent; chroni-cally malnourished—27.5 percent; acutely malnourished—9.0 percent. March 1998: Underweight—22.8 percent; chronically malnourished—26.7 percent; acutely malnourished—9.1 percent.

POSTSCRIPT

page 289 Half of all Iraqi children still malnourished: *Washington Post*, 12/13/98.

Index

171–74, 181–82, 185–90, 214–16,
227, 278–79
Iraqi National Congress (INC) and,
56–57
Kurdistan and, 239–40
as peacemaker between Kurdish factions, 179–80
Chemical weapons, 91, 93. *See also*
Nerve gases
Cheney, Richard, 32
Churchill, Winston, 216
CIA
Chalabi and the INC and, 51–53,
165–67, 171–74, 178, 181–82,
185–90, 190, 214–16, 227, 278–79
covert operations against Saddam
and, 31–32, 34, 37, 45, 165,
166–67, 168–71, 230
decision to send officers to Iraq and,
171–74
economic sanctions against Iraq and,
44, 55–56
Hussein Kamel's defection and, 197
invasion of Kuwait and, 35–36
Iran-Iraq war and, 34–35
Iraqi National Accord and, 174–75,
182–83, 229–30
Iraqi weapons production and, 105
Jordan and, 218–20, 224–25
Kurds and, 176, 248
propaganda used by, 53–55
Saddam's rise to power and, 74–75
terrorism bombings and, 211–14
Unscom inspections and, 98, 100
Clientism, 215
Clinton, Bill
covert actions in Iraq and, 220–21,
226
Deutch's appointment to the CIA
and, 216
economic sanctions and, 165
Kurds and, 164–65, 184, 237, 243
Unscom inspections and, 272–73,
281–85
CNN, 225, 276
Cockburn, Patrick, 225
Cohen, David, 217, 220

Cohen, William, 275, 276
Communist Party, Iraqi, 48, 150, 166
Cosenza, John, 250
Crime, and economic sanctions,
120–21, 127–28, 286
Critchfield, James, 74

al-Dalaimi, Mohammed Mazlum, 192,
255
Dalaim tribe, 149, 167, 192, 255
al-Dawa, 79–80, 254
Defense Intelligence Agency (DIA)
covert actions in Iraq and, 220–21,
244
Iran-Iraq war and, 80
Deutch, John M.
appointment to CIA, 216–18
Chalabi and INC and, 227
covert actions in Iraq and, 230
Devine, Jack, 216
al-Dhour, Iraq, 72, 73, 277–78
al-Dhouri, Izzat Ibrahim, 28, 144
al-Dhouri, Omar, 229
al-Dhouri, Riyadh, 229
al-Dhouri tribe, 144, 149
Dillon, Gary, 282
Diyala province, 132–33
al-Dohra power station, Baghdad, 3, 4,
127
Dohuk, Iraq, 28
Dole, Robert, 245
Dubdub, Mohammed, 152
Duelfer, Charles, 268–69, 276
Dulles, Allen, 67

Economic sanctions. *See* Sanctions
against Iraq
Egypt, 73–74, 153, 284
Ekeus, Rolf, 265
background of, 96–97
collection of evidence and, 108–10
confrontation between Iraqi officials
and, 105–7
departure of, 268, 270
Hussein Kamel and, 197–99, 205, 206
Unscom team and, 97–98, 99–102,
238

Hussein, king of Jordan, 7
  Hussein Kamel's defection to Jordan
    and, 194–96, 210, 218
  Hussein Kamel's return to Iraq and,
    208
  role as counselor to Saddam's family,
    155
  Saddam's assassination attempt
    against Qassim and, 72
Hussein, Ahmed, 162
Hussein, Khalid, 262
Hussein, Qusay (son), 141, 151, 256,
    261
  concealment of secret arsenal and,
    266
  coup attempt against Saddam and,
    228–29, 237
  family background and, 156, 158
  Hussein Kamel's return to Iraq and,
    209
  supervision of security by, 148, 204
Hussein, Raghad (daughter). See
    Kamel, Raghad
Hussein, Rina (daughter). See Kamel,
    Rina
Hussein, Hala (daughter), 143
Hussein, Saddam
  Baath Party and rise of, 70–73
  background of, 67–71
  escape story of, 72–73
  marriage of, 68, 142–43, 153
  U.S. policy and, 218–19
Hussein, Sajida (wife), 151
  Hussein Kamel's return to Iraq and,
    207
  marriage to Saddam, 68, 142–43, 153
  Uday and, 155
Hussein, Uday (son), 140–41, 150–51
  attempted killing of, 251–60
  background of, 142–44
  business dealings of, 161, 162–63,
    193, 225, 253
  education of, 151–52
  Hussein Kamel's killing and, 209
  people's fear of, 141–42, 156–57
  rising power of, 157–63, 192–93
  Saddam's curbing of, 202–3, 204

shooting of Jajo by, 153–55
survival and resurgence of, 261–62

IAEA (International Atomic Energy
    Agency), 86, 100, 103, 282
Ibrahim, Ahmad, 194
Ibrahim, Barzan (half-brother), 68, 77,
    155, 262
  Hussein Kamel and, 146, 147, 202, 207
  nuclear weapons program and, 88, 89
Ibrahim, Hardan, 76, 144
Ibrahim, Sabawi (half-brother), 68,
    144, 261
  attempt on Uday's life and, 254
  supervision of security by, 147–48
  Uday's criticism of, 192
Ibrahim, Watban (half-brother), 68,
    144, 161, 194, 261
  attempt on Uday's life and, 254
  Uday's criticism of, 192, 211
  Uday's shooting of, 202, 203
Ibrahim al-Hassan, 68, 144
Ibrahim family, 143–44, 147
Ickes, Harold, 221
Immigration and Naturalization Ser-
    vice, 248
INC. See Iraqi National Congress
Intelligence agencies, Iraqi
  elimination of potential rivals of Sad-
    dam and, 11
  Iraqi National Accord activities and,
    226–27
  Kurds and, 185–86
  support for uprisings against Saddam
    and, 24–25
Intelligence agencies, U.S.
  attempted killing of Uday and, 255
  Hussein Kamel's defection and, 197,
    220
  individuals fleeing Saddam's regime
    and, 173–74
  invasion of Kuwait and, 35–36
  Jordan and, 218–20
  Kurds and, 36–37, 242–43
  Soviet development of biological
    weapons and, 98–99
  Unscom and, 267–68